THE FOREVER FIX

THE FOREVER FIX

GENE THERAPY AND THE BOY WHO SAVED IT

RICKI LEWIS

ST. MARTIN'S PRESS
NEW YORK

 For information, address St. Martin's Press, 175 Fifth Avenue, New York, N.Y. 10010.

www.stmartins.com

Library of Congress Cataloging-in-Publication Data

Lewis, Ricki.
The forever fix : gene therapy and the boy who saved it / Ricki Lewis.
p. cm.
Includes bibliographical references.
ISBN 978-0-312-68190-6 (hardcover)
ISBN 978-1-4299-4147-1 (e-book)
1. Gene therapy—Popular works. I. Title.
RB155.8.L49 2012
616'.042—dc23

2011038193

First Edition: March 2012

10 9 8 7 6 5 4 3 2 1

For the children . . .

CONTENTS

PART IV BEFORE GENE THERAPY

PART V AFTER GENE THERAPY

PART VI COREY'S STORY

PREFACE

IN THE FALL OF 2008, THE HOT-AIR BALLOON THAT HOVered in the distance along Interstate 76 bore tiger stripes, marking the Philadelphia Zoo below. As it loomed, the tethered balloon provided a reprieve for parents stuck in the ever-present traffic, giving restless and excited children something to focus on other than escaping the car. Once they got to the zoo, youngsters scampered up the path leading to the entrance, seeking the strings that held the giant, colorful balloon aloft.

For the family of three from Upstate New York, the approach on the path to the zoo's entrance was slower. Even if they hadn't been holding hands, it would've been clear that they were a unit, all three with light reddish-brown hair and blue eyes. It was just four days after Corey's gene therapy, and the eight-year-old, still using his silver-tipped cane out of habit, stepped hesitantly, looking down at the few orange and yellow leaves on the path. When the family reached the iron entrance gate, Corey stopped, and like the other kids, couldn't help but gaze up at the enormous balloon. Then he let out a shriek.

"It hurts!" Corey cried, shielding his eyes with his hands.

His parents were momentarily stunned. Then it very slowly dawned on them what was happening. *Could it really be? So soon?* They gently pulled Corey to the side of the pathway so others could pass.

"Are you okay? What do you mean, it hurts your eyes?" asked

Nancy, smoothing his hair, her face a mix of concern and joy. To this day she cannot tell this tale without crying.

"The light! It hurts!" Corey repeated, slowly lowering his hands and squinting. He was terrified because this had never happened to him before. He'd so often stared right into a lit bulb with no reaction at all, even as an infant.

Corey had been born with an inherited condition called Leber congenital amaurosis type 2—LCA2. A single gene had a glitch that prevented his eyes from using vitamin A to send visual signals to his brain. Legally blind, he was headed toward a world of total darkness by early adulthood. But a groundbreaking medical experiment had sent viruses bearing healthy genes into his left eye, with the spectacular results realized not at the hospital or in a lab, but at the zoo—in just days. Gene therapy was sorely in need of good news, for in that same city, nine years earlier almost to the day, an eighteen-year-old had died shortly after such an experiment.

Hands shaking, Ethan fumbled with his cell phone, finding the direct line to Dr. Jean Bennett, impatient as the seconds passed. *What was taking so long? Why didn't she pick up right away?* He nervously shifted his weight from foot to foot as his gaze remained on his son, who now had tears streaking his cheeks. When "Dr. Jean" answered, Ethan could barely get the words out.

"It's Ethan," he stammered. "The sun . . . It hurts his eyes!"

For the very first time, Corey Haas could really see. And in that moment, a biotechnology was reborn.

Four days earlier, Corey had been legally blind, heading toward a word of total darkness.

PART I

THE BEST THAT CAN HAPPEN

The lightning bugs were out tonight, their butts all lit up. Corey could see them!

—Ethan Haas, Facebook, June 28, 2010

1

MEETING COREY

WHEN I MET NINE-YEAR-OLD COREY HAAS, IT WAS hard to believe that just over a year earlier, he had been well on his way to certain blindness. As a geneticist, I'd followed the incredible story of his gene therapy, and as a mom, I was curious about the young man who had made medical history.

On a dazzling Saturday morning in early December 2009, I drove forty-five minutes north to meet Nancy, Ethan, and Corey Haas, who live in the Adirondack Park. This spectacular expanse of nature is not a park in the conventional sense, but a vast swath of New York State peppered with small, older houses and log cabins nestled among towering pines, with the occasional noticeably newer vacation home of a downstater. The meandering Hudson River, only a few hundred feet across this far north, snakes through the Haas family's small town of Hadley, where Corey's parents grew up. The river is only a block away from their Cape Cod home, flooding their basement in bad snow years. I'd been in the area before, so I found it easily, but some of the reporters coming to interview the family over the past month had grown uneasy when the towering mountains blocked even the most powerful cell phone and GPS signals. But they made it—Ethan gives good directions.

Although it took only days for the researchers to realize that Corey's gene therapy had worked, they had waited a year to publish the results, to be sure the effect lasted and hadn't caused any problems. They had learned from past gene therapy fiascoes, and also knew Corey's newfound sight would be big news. And they were right, with *Good Morning America* setting the slightly hyped tone by describing "revolutionary surgery that could give the gift of sight to the blind." When the research paper appeared in *The Lancet* in November, the Haas family traveled to New York City for the very first time, where they made the rounds of the major TV news programs. Despite the intense interest, though, only a few journalists ventured north to meet Corey on his own turf.

The morning of my visit, the scent of the season's first snowfall was in the air. In the Adirondacks, you don't need a weatherman to know which way the wind blows, to quote Bob Dylan. You just have to watch the squirrels. I had to hopscotch around them as I walked up to the Haases' front door, my hands laden with my laptop, books, and a large box. The squirrels scurried everywhere, grabbing and stashing acorns in the heart-shaped crevices at the bases of the tree trunks, occasionally dashing the wrong way and bashing into the large picture window in the living room that reached almost to ground level. The squirrels knew snow was coming.

When Corey opened the door, our eyes went first to the scampering squirrels, then to each other. He gave me a huge grin, his blue eyes dancing behind thick glasses. He took my leg and guided me to a coffee table, where I dropped my armload. As Ethan and Nancy introduced themselves, Corey began rummaging around the rocks and fossils in the box, a subset of the collection I'd begun at his age. Knowing he'd only recently been legally blind, I knelt down and tried to show him how to feel the indentations of the trilobites and brachiopods. But he proudly refused all assistance, peering closely at each fossil, as I had at his age, slowly turning each one to see what different angles might reveal.

I looked around. The décor was a welcoming blue and white, the rooms unusually tidy for the home of a small boy. His toys were neatly aligned on low shelves and stacked in corners, without the clutter that continually grows in my own home. But perhaps this was a habit from the time when Corey had to memorize where every single item in a room was in order to move about. Above the couch where I sat and typed as the family talked, a shelf held a dozen of Nancy's 110-plus dolls. These were the special ones, the others inhabiting closets and a storage room, packed in with Nancy's scrapbook collection. The other walls held what can only be described as a gallery of Coreys. My favorite was a photo of Corey at about one year old, with a broad smile and thick glasses, "Corey" stitched in needlepoint below, framed in red. The earliest photo on the walls, taken when Nancy and Ethan had yet to learn that Corey had inherited an eye disease, showed baby Corey stuffed into railroad overalls, a yellow shirt, and white socks, looking off to the side as if he'd rather be anywhere else. A wall in the dining room had the words *live*, *laugh*, and *love* spelled out in wood, with three photos beneath each—all of the resident celebrity, Corey.

As Ethan talked and sorted copies of medical reports to give me, Corey jumped up every few minutes to show his mother a rock. I glanced at them, trying, but failing, to imagine this energizer bunny of a boy stumbling around the same living room, crashing into the furniture like the squirrels outside careening from tree to tree. But that day he rocketed around the cozy living room with ease. He showed me his handheld Nintendo and how to extract a fart noise from a container of Silly Putty–like goop, and then he asked me to name the minerals in his rocks. Periodically he'd demonstrate the hardest part of the gene therapy for him—lying ramrod still on his back. I noted subtle signs of what the family had been through. Once in a while Corey would lose his footing briefly, and Nancy's arm would instantaneously zap out to steady him, sometimes hugging him close. He didn't squirm away. Corey's independence was still new to all three of them. For years he could see only shadows, and in dim light, nothing at all.

On September 20, 2008, two days before Corey's eighth birthday, and fifteen months before I met him, the Haas family had traveled to Children's Hospital of Philadelphia—CHOP. That week, Corey underwent a battery of what had become familiar tests, plus a few new ones, and on Thursday morning was finally prepped for the procedure. Then, as Corey lay anesthetized, the eye surgeon Al

Maguire snipped open a tiny flap in the left eye, the worse one, and carefully injected 48 billion doctored viruses into a tiny space just above a thin layer of colored cells that resembled patterned bathroom tile. This space and the layer of cells hug the rods and cones at the back of the eye. Corey's colored cells weren't doing their job of nourishing the precious rods and cones, which are called photoreceptors because they capture incoming light. His rods and cones had been slowly starving for years, gradually becoming too weak to send light signals to his brain to paint an image. As a result, before Corey lay down on the operating table, he saw only blurry, dark shapes. But that would change—sooner than anyone had even dared imagine.

2

THE ROAD TO A DIAGNOSIS

COREY HAAS WAS BORN ON SEPTEMBER 22, 2000. AFTER an uneventful pregnancy, Nancy gave birth to the blond, blue-eyed baby, who weighed in at a respectable nine pounds.

For the first few months, all was perfect and he hit his developmental milestones ahead of schedule. "Corey was a happy baby. But the babysitter noticed that he wasn't reaching for things right in front of him, like other babies do," recalled Nancy. She worked in an office nearby, while Ethan commuted more than an hour each way to and from his job at International Paper, so they often weren't with their son during the day. Corey's parents and the babysitter wondered about Corey's peculiar habit of staring, enthralled, at lit bulbs, for far longer than anyone else would or could. When the babysitter commented that Corey never made eye contact with her when drinking his bottle, Nancy realized that she'd been aware of this too, but thought he was just too young to focus.

As the weeks went on, the new mother felt an ember of fear whenever she'd call her son's name and he'd turn, but just stare in her general direction; she didn't feel as if he was *seeing* her. Then Nancy noticed that he played only with toys that were lit up or that were lying in a patch of sunlight on the carpet. Something was wrong with Corey.

By six months of age, the problem became more pronounced. Not only couldn't Corey focus, but now his eyes wandered, flicking back and forth, his left eye worse than his right. This painless but unsettling condition, which the Haas family later learned is called nystagmus, is common in people who have albinism. Perhaps because Corey's hair and skin had color, the local pediatrician reassured the concerned parents that albinism didn't seem likely, and that the boy would probably outgrow the troubling symptoms.

Corey's local pediatrician apparently made no notes that indicated he suspected a genetic disease, and this is understandable; such conditions usually bring a swath of signs and symptoms, such as developmental delay, unusual facial features, defects in major organs, or constant colds from a suppressed immune system. Corey was a chubby, vibrant picture of health, adorable and active. But the pediatrician, noting at the six-month exam Corey's occasionally crossed eyes, was sufficiently concerned to refer him to the ophthalmologist Gregory Pinto in nearby Saratoga Springs, the first in a series of eye specialists who would gradually zero in on the diagnosis. Nancy and Ethan were concerned, but not alarmed. Perhaps his visual system was just developing a little slowly—not all babies focus on their parent's faces in the first weeks of life. Corey's wandering eyes seemed like something that would be simple to fix, if the condition didn't clear up on its own.

When Dr. Pinto first saw Corey, the boy was seven months old. Corey's eyes had all the right parts, but the doctor noted that the back of his eyes had unusually pale areas, "blond fundi" in the medical lingo. The most telling of the doctor's observations was Corey's fascination with lights. Dr. Pinto had seen that before, in people of limited sight trying to "self-stimulate," he told Nancy.

Except for his eyesight, Corey seemed fine, eager to go from immobile infant to active toddler. He loved to explore. At eight months, no longer content to sit and crawl, Corey was already hauling himself upright, but he was clumsy. He'd bump into things, especially when the room was dimly lit. And Corey continued to stare at lights, mesmerized.

Like a seedling seeking the sun, Corey continued to pull himself up and grab toward the lamps in the small living room. Soon he was cruising and then walking. Once he could no longer feel the surface of the shaggy carpet beneath him with his hands and lower legs, his internal feedback system was gone, and the bumping into things worsened. Yet in other ways he was right on track. He could utter a few words and even draw simple shapes. He was learning to compensate with his better-functioning senses.

At the next eye exam, Dr. Pinto picked up on a change for the worse. "Corey had started to show signs of visual difficulty. Sometimes he would crawl right into a table leg. He didn't follow objects as well as he previously had. His eyes now crossed slightly and, most significantly, he was quite nearsighted," recalls the ophthalmologist. The doctor was concerned at the rapid deterioration of Corey's vision and sent him to a pediatric ophthalmology subspecialist, Dr. John Simon, an Albany physician who had a satellite office near the Haases' home. He confirmed Dr. Pinto's findings, and prescribed Corey's first glasses when the boy was ten months old. At a follow-up visit near the end of 2001, Dr. Simon discovered some new signs: Corey's left eye turned inward, and his irises let light through in spots, as if pierced with tiny holes. But Corey was still too young for anyone to really tell what was happening, other than extreme nearsightedness. Many kids simply outgrew early visual problems. It was too early for a definitive diagnosis, but in the meantime, the glasses would help.

Corey continued to develop at a normal pace, but his vision worsened, despite doubling the correction in his glasses. At his last visit with Dr. Pinto, Corey was in the throes of the "terrible twos." He wouldn't keep still long enough for a complete retinal exam, but the doctor saw enough to realize that Corey was *trying* to look at objects and identify them, but he couldn't really *see* them. And he still stared, captivated, at the lights.

In early 2003 came the first critical connection that would catapult Corey toward gene therapy. Although he did not yet have a diagnosis, Corey saw so poorly that he was eligible for the New York State Early Intervention program, which assists young children who have disabilities or developmental delay. One day, the early-intervention provider, visiting the Haas home, mentioned in passing that she had been talking to another parent whose child had a similar visual problem.

"Where did the parent take her child?" Nancy asked.

Boston Children's Hospital is about a five-hour drive from the southern tier of the Adirondacks, counting bathroom stops and the inevitable traffic on the Mass Pike. In January, an ice storm or blizzard can derail travel plans, as can the squalls that materialize out of nowhere along the barren stretch of interstate in the Berkshires, whipping up ephemeral mini-tornadoes of blinding snow. But luck was with the Haas family on a January day in 2003. The air was clear and crisp, and they made it to Corey's first appointment at Boston Children's Hospital in record time.

They instantly liked Anne Fulton, an ophthalmologist specializing in diseases of the retina in young children. Her shock of thick white-gray hair looked a little like Corey's blond mop. Dr. Fulton listened intently to Nancy and Ethan and wrote in the medical chart, "Corey's visual behaviors are described as seeing well in good light. In low light, he crashes into the furniture, and they have wondered if this might be because he is looking at the lamp instead of where he is going. He watches the TV at an angle."

The next day, Corey was anesthetized to keep him still and given eyedrops to dilate his pupils, to do an electroretinogram, or ERG. This test shows the retinas' responses to flashes of light, and is normally a curve that dips and then sharply rises.

Corey's ERG was as flat as an Iowa cornfield.

Ophthalmologists knew that a flat ERG is a definitive sign of a certain class of retinal diseases, but since Dr. Fulton rarely had seen such a profound lack of a response, she double-checked her equipment. It was fine. The problem *was* with Corey's eyes. Light energy wasn't getting to his brain.

Once Corey awakened from the anesthetic, Dr. Fulton donned a device resembling a miner's helmet and gently peered at each of the boy's retinas, under light so intense it often made children (and some adults) scream. Not Corey. If he hadn't been so fascinated with all the paraphernalia, and so oblivious to the penetrating light that was much brighter than he was used to, it might have been harder to keep him still. But Dr. Fulton got a good look. In each eye, she saw that the macula, the pale area near the point that provides the sharpest vision, was *too* pale. It was an ominous sign.

Dr. Fulton's tentative diagnosis: albinism. Corey's stumbling in dim light suggested night blindness, but the family's pale coloring and the observation that both Nancy's and Corey's irises let in light through tiny holes suggested albinism. Perhaps Nancy actually had a very mild case of albinism, and Ethan was a carrier of the same type, and their individual mutations combined in a way that more severely affected their son's eyes. The various types of albinism are recessive, passed from unaffected carrier parents, so the fact that neither Ethan nor Nancy knew of any affected relatives didn't matter. Albinism could be passed, silently, for generations. Corey had already tested negative for the most common type of albinism, but maybe the Haas family had a rare or even unique form of the condition that affected mostly the eyes.

Eight months later, in the fall of 2003, Corey turned three and started preschool. His world now consisted of shadows even when he was in a normally lit room, and in a brightly lit room he had extreme tunnel vision. Today he acts out this early memory, tilting his head to show what he once had to do to position an object in the center of his shrinking visual field.

That October, Corey returned to Boston Children's Hospital for a more complete workup. Wen-Hann Tan, a pediatrician and geneticist, examined the squirming toddler, checking for hints of "dysmorphology"—known informally in some genetics circles as an exam for an FLK, or funny-looking kid. People with genetic syndromes often have unusual facial features that alone might simply seem quirky, but taken together may point a perceptive clinician to check particular genes and chromosomes. Corey sat still while Dr. Tan meticulously measured the dimensions of his ears, nose, and jaw; opened his mouth wide to scrutinize his palate; noted the spacing and slant of his eyes; and measured the space between his upper lip and nose. Corey's face was cherubic, bearing not a hint of anything awry.

Still not ready to abandon the possibility of albinism, Dr. Fulton asked the attending physician in genetics and metabolism at the hospital, David Harris, to test Corey, Nancy, and Ethan for two rare types of the disorder. In albinism, an enzyme deficiency blocks production of melanin pigment. Without the pigment, the eyes can't form good images, and the optic nerve pathways to the brain don't develop normally. The visual field shrinks, and the eyes jiggle back and forth. Again, Corey's results were normal. His eyes weren't normal, but it wasn't because of albinism. What was wrong?

With the more common explanation, albinism, now ruled out, retinal degeneration was rising to the top of the list of possible diagnoses. Corey's peripheral light-sensing cells, his rods, and possibly his central color-vision cones, were either wasting away or ignoring incoming signals. This was more dire than albinism.

Corey, Ethan, and Nancy returned to Boston Children's Hospital in May 2004. Dr. Fulton's notes from that visit chronicle a growing boy, interspersing technical terms such as *diopter* and *esotropia* with *wiggly.* "Corey is active. He is an explorer." Her analysis of the specific visual deficits over time now led her to suggest a condition called Leber congenital amaurosis (LCA). Back in 1998, she'd been part of the team that identified the gene that would turn out to be behind Corey's condition. Now she wrote in his chart, "Corey's visual acuities are too good, or much better than in many children with LCA, for us to think of LCA in a conventional way."

It was the first mention of what would ultimately be the correct answer.

3

WHAT'S WRONG WITH COREY?

UNDERSTANDING WHAT GOES WRONG IN LEBER CONgenital amaurosis requires a trip through an eyeball, from the pupil, where light enters, to the back of the eye. Here, the retina is the layer of the eye's wall that includes the photoreceptor cells—the rods and cones—that capture light energy and change it into the electrical language of the nervous system. The rod cells provide black-and-white vision and detect motion, and the cone cells send signals for color. The retina also has cell layers that transmit the light signals to the optic nerve, which sends the information to the part of the brain that interprets the input as a visual image. The comparison of the human eye to an old-fashioned camera is apt—the back of the retina is like a sheet of photographic film.

At first Corey's night blindness suggested a problem with his rod cells. Each eye has 100 million of these long, skinny cells, and each has about two thousand translucent discs that fold inward from the surrounding cell membrane, making the rod look a little like an electric toothbrush. The aligned discs resemble toothbrush bristles at one end, and a neural connection at the other end of the cell that goes to the brain corresponds to the part of the toothbrush that plugs into a power outlet.

Embedded in the rod's folded discs are many molecules of a pigment called rhodopsin, which actually provides vision. Each rhodopsin molecule is built of a protein part called opsin and another, smaller part made from vitamin A, called retinal. A flash of light lasting mere trillionths of a second changes the shape of the retinal, which in turn changes the shape of the opsin. The change in opsin triggers chemical reactions that signal the nearby optic nerve, which stimulates the visual cortex in the brain. In this way, each of the 100 million rods and 3 million cones of a human eye contributes a tiny glimpse of a scene, which the brain then integrates into an image.

To see the world as a continuous panorama, rather than a series of disconnected snapshots, rhodopsin must quickly reform after it changes in response to light. In dim light this happens slowly, and the rhodopsin is recycled inside the eye. But in very bright light, rhodopsin contorts too fast to fully recover. This is why we are temporarily blinded when walking out of a dark theater, to which our rhodopsin has adapted, into bright sunshine. It is also why we tell children to eat their carrots, because vitamin A deficiency causes night blindness. Cones work in a similar way, but instead of rhodopsin, they use three other visual pigments that are sensitive to different wavelengths of light, and interpret the hues of red, green, and blue that color our world. Mammals other than humans and our primate cousins have only two types of cones, which restricts the color palette available to them.

If Corey indeed had a form of Leber congenital amaurosis, the origin of his difficulty seeing wasn't in his rods and cones, but in a layer of cells next to them called the retinal pigment epithelium. The RPE is a caretaker of sorts for the rods and cones, removing wastes while absorbing stray light rays that might otherwise bounce around the eyeball, creating meaningless flashes. The RPE's most important job is to store vitamin A. It uses a protein, called RPE65, to activate the vitamin, forming the retinal essential for black-and-white vision. This would turn out to be Corey's precise problem: his eyes can't make RPE65. The condition is inherited, because genes tell cells how to make specific proteins. Without normal RPE65, nearly all of Corey's rods and cones would shrivel away to nothing. By age forty—and that was being extremely optimistic—he would be completely blind, and he would be legally and functionally blind far earlier. Yet because photoreceptors are so abundant that even the blindest of the blind still harbor a few of the cells, and the fact that Corey's young age meant that many had not yet been starved, this particular form of LCA was an ideal candidate for gene therapy. And the younger the patient, the more likely it was to work. But LCA comes in at least eighteen different forms, each caused by a different abnormal gene. A genetic test would be necessary to turn Dr. Fulton's notation in the medical record into a definitive diagnosis.

It's little wonder that it took years to identify the cause of Corey's disappearing vision, since mutations in more than 180 genes can harm the retina. A common form of hereditary blindness is retinitis pigmentosa (RP), in which the photoreceptors themselves, especially the rods, shrink and then die. Symptoms of RP do not usually begin until early adulthood. Some researchers classify LCA as a form of RP, even though LCA is an indirect assault on the photoreceptors. However the classification works, the subtypes of LCA account for about 8 percent of inherited retinal dystrophies, but for 20 percent of children in schools for the blind, due to the early onset and severity.

LCA was recognized long before researchers knew that it stems from mutations in any of eighteen-and-counting genes. A German ophthalmologist, Theodore Leber, first wrote about the familial form in 1869. However, early reports indicate that not all cases are congenital—some are environmental. A case report from 1886 described a young woman living in Savannah, Georgia, who developed the condition after taking too much quinine to treat malaria. When she awoke from a coma, she couldn't tell light from dark. The report is eerily like a description of Corey as a toddler attempting to navigate his living room. "She cannot see to go alone, but with care she can distinguish large colored objects in her room, and with some hesitation, constantly moving her eyes as she does so. She can count fingers at four feet, but she cannot make out a letter." Luckily for her, the effects of quinine-induced amaurosis were temporary.

Amaurosis means "loss of sight." One type of amaurosis, called "fugax," stems from a fleeting circulatory disturbance in the eye, causing visual loss that lasts only a minute. It can be a warning sign of an impending stroke. Another form of amaurosis affects ruminants—cattle, sheep, goats, deer, and camels. They can develop the condition from vitamin B_1 deficiency, which happens in two ways. One is when they eat certain ferns that disable a key enzyme. The second way is when there is a change in the bacterial populations in one of the four stomachs of such a creature, which prompts an explosive release of the foul anal gas hydrogen sulfide.

Corey returned to Boston Children's Hospital on his fourth birthday. His condition was rapidly deteriorating, his tunnel of vision narrowing. Examining the backs of his eyes, Dr. Fulton found that each macula had developed a dark dimpled area in the center of the unusual pale circle. One bad eye could be a congenital fluke; two meant an inherited problem because the simultaneous occurrence of two rare events most likely stemmed from a fundamental error. Was it retinitis pigmentosa, perhaps the type that a mother who is a carrier passes to her son? But Nancy's genetic test for it was normal.

Even as recently as 2004, genetic testing was painstakingly slow, as the analysis of the human genome sequence continued. Researchers had determined the entire DNA sequence in 2000, publishing final results by 2003. And so doctors would still try one or a few genetic tests at a time, those that seemed most appropriate as a child's symptoms unfolded and simply because not many tests existed. Today not only are panels of genetic tests available for many diseases, but genome sequencing can identify mutations apparently unique to a particular family—this indeed happened for two LCA families who are friends of the Haases' in the summer of 2011.

But for Corey, genetic testing led in a logical, if slow, fashion toward his diagnosis.

During the next visit to Boston, at the end of January 2005, it was Nancy and Ethan's turn for ERGs, an experience that they do not remember with fondness because probes must be affixed directly to the eyeballs. Their scans showed low retinal function, but not low enough to affect their vision. This was a key clue. They could be carriers of the same form of inherited retinal disease, with symptoms so mild that they aren't noticed. With these ERG results, the next step was to test the family's DNA for LCA, because Corey's flat ERG ruled out other inherited eye conditions. But even that wasn't straightforward, because new genes were being discovered all the time. Ethan remembers this visit as a turning point. "Dr. Fulton asked us to reconsider the genetics, because a number of genes had been identified in research." New genetic tests were becoming available.

Meanwhile, clinical clues also helped the doctors rule out certain other inherited eye diseases. They'd already eliminated conditions that might explain a child who sees better in the dark, the opposite of Corey (achromatopsia, cone dystrophy, or Stargardt disease), and concentrated on explanations for a child who sees better in the light (retinitis pigmentosa, congenital stationary night blindness, or some forms of LCA). Specialists had also crossed off the diagnosis list syndromes that include night blindness: Corey didn't have the extra fingers or toes of Bardet-Biedl syndrome, the brain degeneration of Batten disease, the deafness of Usher syndrome, or the paleness of albinism.

A general diagnosis of LCA is possible based on symptoms and test results such as the ERG, but genetic testing explains variations by identifying disease subtypes. Physicians and researchers use the different gene symbols, italicized abbreviations usually denoting the affected protein, in a language of sorts with which parents rapidly become quite fluent. Corey's *RPE65* form of LCA has mild improvement in early childhood, and then a course downward later in life, as do the *CRB1* and *LRAT* subtypes. Children whose vision progressively but slowly declines could have *AILP1* or *RPERIP1*, and infants born into total darkness could have the more common forms of LCA caused by mutation in the *CEP290* gene or *GUCY2D* gene. Another diagnostic clue is that children with some types of LCA have a characteristic eye-poking behavior called the "oculodigital sign." The appearance of the back of the eye holds clues, too. It is "blond" in some forms, like Corey's, or may have a bull's-eye-like center in another type. LCA comes in so many varieties because the functioning of the rods and cones is so complex that many proteins are involved in vision, and therefore many genes. Dr. Fulton's careful medical record would help in selecting the specific genetic tests that would most likely finally answer the question "What's wrong with Corey?"

Dr. Harris, the geneticist at Boston Children's Hospital, sent blood from the Haas family to the John and Marcia Carver Nonprofit Genetic Testing Laboratory. Unlike other facilities whose offerings cover many body parts, the Carver lab specializes in the eye. They test for most, but not all, of the genes known to cause LCA. "Looking for LCA genes in the human genome is not like looking for a needle in a haystack, but like looking for a silver needle in a haystack of steel needles," explained Edwin Stone, MD, PhD, director of the lab and professor of ophthalmology and visual sciences at the University of Iowa in nearby Iowa City. He was speaking at a meeting for LCA families on the University of Pennsylvania campus on a midsummer weekend in 2010.

The lab is not-for-profit, but testing usually isn't free because sequencing entire genes, which may be necessary to spot a rare mutation, is expensive. Finding a common LCA mutation might cost only $700, but if it's rare, the charge goes up by $600. The price tag for an accelerated, six-week turnaround is $2,500. About a quarter of families who submit blood samples to the Carver lab don't get an answer, and end up paying the fee to eliminate possibilities. They then seek out other facilities that test for additional mutations, such as the DNA Diagnostic Lab at the University of Colorado in Denver.

If answers keep coming back negative, families have their entire "exomes" (the part of the genome that encodes protein) sequenced. A decade from now, once researchers have identified all human genes and their variants, finding a mutation that causes a specific disease will be as easy as typing a book title into Amazon.com.

To speed genetic-testing identification of the estimated three thousand people who have LCA, the Carver lab is running Project 3000. Wyc Grousbeck, owner of the Boston Celtics, and Derrek Lee, first baseman of the Chicago Cubs, started the project in 2005, after each had a child diagnosed with the disorder. (Derrek Lee's daughter turned out to have a viral infection and not LCA, but he's stayed involved.) If a family lacks health insurance to cover testing, donations may defray costs.

I was initially put off by denials of my requests to visit the Carver lab or even to speak to Dr. Stone, but then I learned that he works nearly round-the-clock, tracking down genes, developing new tests, and talking to frantic parents at all hours. I found myself next to him at the LCA family meeting, and tried to explain that I needed to talk to him for this book. Just then, the speaker announced that Dr. Stone would be available to meet with parents for ten-minute time slots and pointed him out, and he was instantly mobbed. He smiled at me as parents lined up, and he shrugged. "See the problem?"

Fortunately, Corey's LCA mutation is so common that eight labs test for it, including Carver. The family finally traveled to Boston on July 5, 2006, to learn the results from Dr. Fulton.

Corey was now nearing his sixth birthday. He was making excellent progress in learning Braille, but he needed so much paraphernalia in his classroom in order to see that he and his stuff took up two seats. The teacher put him in the back of the room because his equipment distracted the other kids if he sat in the front. He remembers the isolation well. "I had a Clarity DeskMate, which is like a little computer monitor screen with an arm with a camera attached that let me see the whiteboard. I had to sit in the back." To read at his seat, Corey needed a magnifying glass and a 100-watt bulb, and still he had to keep his face right next to the print.

"If Corey was in a typical classroom, where the kids would find the light bright enough to read, he could not. His pupils would dilate as far as they could, as if they were trying to pull every photon out of the room," explained Dr. Jean.

Dr. Fulton told Nancy and Ethan that Corey had inherited a different mutation, but in the same gene, from each of them. This meant they weren't distant cousins, which is always a possibility with an extremely rare genetic disease. Relatives can inherit the same mutation from a shared ancestor, such as a common great-grandparent. Ethan's genetic quirk is localized: one DNA base subs for another, like a one-letter typo. Nancy's class of mutation, called "nonsense," prematurely halts her cells' reading of the gene, releasing just a nubbin of RPE protein, if any at all. ("I always thought you were full of nonsense!" joked Ethan.) As carriers, Nancy and Ethan could see because each has a working copy of the gene. Team their two mutations in the eyes of a child, however, and his cells can't make the needed RPE65 protein at all. Corey's rods were starving—and soon his cones would be, too.

Ethan and Nancy couldn't recall much of the Genetics 101 discussion they had with Dr. Fulton in Boston on that July day in 2006. Despite the years of doctor visits, and the mounting clues that Corey had inherited something, they were still in shock. But they managed to hear enough to realize that each child of theirs stood a 25 percent chance of having Corey's disease. One single word stopped them from absorbing further genetic details. "I remember when I first heard the word *blind*. I thought immediately of all Corey wouldn't be able to do," says Nancy years later, still tearing up at the memory.

Dr. Fulton's recommendations were bleak. Corey should continue his vision-support services at school and wear his glasses, she said kindly, and continue his efforts to learn Braille. He needn't return to Boston for another year, because there was nothing more they could do. She did, however, mention ongoing research, but that flew right past the distraught parents trying to comprehend that their child, their only child, was going blind from something that they had given him. The medical report for that visit ends with the prophetic words, "It is the RPE65 type of Leber congenital amaurosis that recently appears to have responded to some experimental treatment. This is a hopeful beginning." Indeed it was—but Corey's terrified parents didn't yet know it.

What they *did* know, finally, was the enemy: a specific mutation in a specific gene. Corey hadn't overdosed on malaria meds, or suffered a stroke, nor was he a flatulent camel. He was, however, a zebra.

Fledgling medical students learn right away the mantra "When you hear hoofbeats, think horses, not zebras." It means suspect first the most common explanation for a patient's symptoms. Round up the usual suspects. In Corey's case the horses were night blindness and albinism, then retinitis pigmentosa. Medical geneticists, however, deal almost exclusively with zebras, most of which are caused by mutations in single genes. Even the more familiar among these, such as cystic fibrosis and sickle cell disease, are rare compared with conditions that reflect more of an environmental input, such as the common forms of heart disease or emphysema. But understanding how the rare genetic conditions happen can often explain more common ills. For example, the statin drugs that millions of people take to lower cholesterol were developed based on studying the one-in-a-million children who died of familial hypercholesterolemia, their cholesterol so high that it collected in waxy yellow clumps behind their knees and elbows. Understanding how the mutant gene revved up the body's own cholesterol production suggested ways to induce the opposite effect.

From finally putting a name to their son's blindness until the next appointment in Boston a year later, Ethan and Nancy navigated a personal hell of helplessness and despair as Corey's visual world continued to close in on him. They also struggled with the special guilt that comes with an inherited illness. "You think, how on earth could this happen to your son? How did I cause this? Then, how am I going to fix it?" recalls Ethan, his and Nancy's anguish easily reawakened. They'd decided not to risk having other children, and were steeling themselves for raising a blind child.

A year later, at the appointment at Boston Children's Hospital on June 28, 2007, everything changed. Dr. Fulton wrote in the record, "a new experimental treatment of gene replacement is planned. We are told that the youngest patients will be considered for this treatment is 8 years, an age that Corey is approaching." She told Nancy and Ethan to bring Corey back in a year, and then she called Dr. Jean Bennett, at CHOP.

But Corey never returned to Boston. Instead, he had gene therapy.

Despite Corey's misfortune in the game of genetic roulette, he was very lucky to be in the right place at the right time. His diagnosis, although it took years of detours, came just as researchers were about to launch a clinical trial to test a gene therapy for not only his rare type of visual loss, but for his *subtype*. The goal: give Corey working copies of the errant gene.

Like most people, Ethan and Nancy Haas had never heard of gene therapy. They had no way to know that Dr. Fulton's note in the medical record would lead to Corey's participating in an experiment that would land him on radio and TV and in headlines around the world, even to being named a "breakthrough of the year" in *Science* magazine. They couldn't know that one day Dr. Francis Collins, the director of the National Institutes of Health, would show the U.S. Congress a video of Corey navigating an obstacle course, stumbling with his treated eye patched, then zipping through in seconds with his untreated eye patched. Dr. Collins, grinning before Congress, shouted, "Go, Corey, go!" as the audience watched the video.

Corey's newfound vision has had a galvanizing effect on the field of gene therapy for several reasons. His plight was easy to relate to—everyone can imagine what it must be like to be so young in a darkening world. A bright, charming child with rosy cheeks and a wide grin, Corey has never fit the image of a person with a genetic disease. And the gene therapy clearly worked. But to understand how Corey figuratively, and perhaps literally, *saved* gene therapy, his success must be considered against the backdrop of a biotechnology derailed, seemingly doomed, by failure and tragedy.

Earlier gene therapy attempts, also on children but for different and more life-threatening conditions, led to questionable or only incremental or transient improvements. For certain inherited brain diseases, gene therapy only slowed the deterioration, extending life but leaving teens with lingering and profound disabilities. And then, in 1999, in another late September in Philadelphia, involving one of the same researchers, an eighteen-year-old had gene therapy for a liver disease.

Jesse Gelsinger was the first, but not the last, to die from gene therapy.

PART II

THE WORST THAT CAN HAPPEN

My son died in a science experiment.

—Paul Gelsinger

4

THE BREAKTHROUGH MYTH

Corey Haas is joining others who have made medical history by being among the first to try a new type of treatment. All of these individuals contribute knowledge that advances medical technology, sometimes by showing when an approach isn't safe or doesn't work, sometimes by showing when it does. Some volunteers, like Corey, improve. And a very few, like eighteen-year-old Jesse Gelsinger, die not of their disease, but from the attempt to cure it. Jesse Gelsinger's tragedy in 1999 stalled the field of gene therapy, halting clinical trials as researchers worked to reevaluate how to best introduce new genes into the body. A look into the history of medical experimentation and clinical advancement adds perspective to Corey's success and Jesse's sacrifice.

Most new medical treatments and technologies are about a half century in the making. Theodore Friedmann, one of the founders of gene therapy and director of the Gene Therapy Laboratory at the University of California, San Diego, offers two examples. "The first bone marrow transplant was done in 1957. It took twenty years before the survival rate crept above one percent." The survival rate improved when immunity-suppressing drugs came on the scene in 1978, and when a dozen years later other drugs became available that boost the number of bone marrow stem cells entering the bloodstream.

Cancer chemotherapy is Friedmann's second example of a medical technology slow to mature. The first clues that some cancers could even be treated came from post–World War I autopsies that revealed too few white blood cells in people poisoned with mustard gas. Might lowering the white blood cell count help people with leukemia, who make too many such cells? Could a toxin in one situation be an effective treatment in another? In 1948 Sidney Farber, a pathologist at Harvard Medical School, discovered the compounds aminopterin and methotrexate, which became the very first chemotherapies. They were used to treat acute leukemia in children. "At first the cure rate was very low. By the 1980s it was sixty to seventy percent, and now it is eighty-five to ninety percent. It took three to four decades to tweak the delivery system, refine the drugs, and add radiation. Each decade, the survival rate jumped ten percent. Cancer chemotherapy is the best example of how we learn from problems," says Friedmann.

Long before doctors studied casualties or chemists probed nature's medicine chest, people must have tested new pharmaceuticals by trial and error. A group of long-ago hunter-gatherers might have come upon an unfamiliar plant, and wanting to know if it had any valuable effects, fed it to a particular individual. If Gork died or became ill, the potential elixir was discarded. But if the people had chanced upon a soothing or healing natural product, its use was continued and refined.

The search for cures became more methodical as healers experimented on people to test new treatments. In 1796, young James Phipps tended the garden of Edward Jenner, an English physician with an interest in infectious disease. Like Corey, James was eight years old when the doctor tried something on him that would become the smallpox vaccine, which ultimately rid the world of this extremely disabling disease. At the time, smallpox killed a third of its victims, and 80 percent of infected children. Painful blisters covered the body and blindness was a frequent complication. Each year, millions died.

For many years in many parts of the world, people had been inoculated against smallpox. This entailed transferring a scraping from a sufferer's skin lesion to a scratch on a healthy person, which was more likely to produce a mild case than a severe one. It was also well known that dairymaids did not contract smallpox. Jenner, after hearing a milkmaid say, "I shall never have smallpox for I have had cowpox," put these facts together to hypothesize that scratching the skin of a healthy person with drippings from the much milder cowpox could protect against smallpox. So he scratched the skin on young James's arm and rubbed into it material from a cowpox pustule found on the hand of a dairymaid, Sarah Nelmes. For a few days afterward, James had only a mild headache, chills, and poor appetite.

James received his first "vaccination," consisting of cowpox-carrying pus, on May 14. The test came on July 1, when Jenner rubbed material from a smallpox pustule into a scratch on James's arm. The boy didn't develop smallpox then, or at any of several other times when the doctor repeated the smallpox exposure. Nor did James pass any pox to the two children who slept with him. He lived, in fact, until the then-great age of sixty-six. It's unlikely that James knew he was in an experiment. Dr. Jenner cared for him and his family for many years. Although Jenner's initial rather verbose publication on his experiment garnered little attention, he distributed his vaccine to many physicians, and as they treated their patients, the efficacy soon became obvious. By 1800, many European nations were routinely vaccinating their populations, and use began in the United States.

At the turn of the nineteenth century, experiments on human volunteers led to a vaccine for yellow fever. At its worst, yellow fever began with chills, vanishing appetite, and then an excruciating headache as fever soared. The pulse slowed, skin paled, blood sugar plummeted, and the abdomen clenched in agonizing cramps. After a few hours of reprieve, symptoms returned with a vengeance, progressing to convulsions, delirium, and copious vomit turned a tarry black from blood. At the end, as organs shut down, the failing liver released torrents of bile, turning the skin and whites of the eyes the vivid yellow that gave the disease its name. Yellow fever was very often rapidly fatal.

In 1900 in Havana, Cuba, Walter Reed's Yellow Fever Board designed and carried out experiments to close in on the cause of the horrific disease. At Camp Lazear—named for one of Reed's closest associates, who'd died of the fever—volunteers lived in one of two experimental areas for days at a time. Building #1 held the clothing of recent fever victims, soiled with all of their secretions, to test the hypothesis that filth and tainted body fluids passed the sickness. It didn't. Building #2 exposed volunteers to mosquitoes that had fed on fever victims. Depending upon how recently the insects had drunk their last "blood meals," some of those men became ill. The experiments implicated the mosquitoes.

By the second half of the twentieth century, many once-common infectious diseases had become rare thanks to improved hygiene, vaccines and antibiotics. Medical research shifted focus to the new most common killer, heart disease. In the early 1980s, two groundbreaking experiments captivated the public, both conducted on human subjects. The retired dentist Barney Clark and the newborn Baby Fae were both near death when they underwent daring new procedures to replace their failing hearts. Unlike a slimy liver or a glistening ribbon of intestine, a heart symbolizes life to the average person, even if doctors distinguish life from death by brain activity. The intense media coverage of the older man and infant receiving new hearts vividly brought the issue of informed consent for medical experimentation to the nightly news.

On December 2, 1982, at the University of Utah in Salt Lake City, Barney Clark received what sounded like a prop from a 1950s-era science fiction film—the Jarvik 7 heart. His own organ was close to giving out from idiopathic cardiomyopathy, a catchall phrase for a heart breaking down for no apparent reason. The Jarvik 7's earlier incarnations had been extensively tested in animals, mostly dogs. The artificial heart was the brainchild of Willem Kolff, MD, who in 1957 had famously kept a dog alive with his prototype heart for ninety minutes, at the Cleveland Clinic. Dr. Kolff moved to the University of Utah, where later generations of devices would take the names of the researchers who perfected them. The graduate student

Robert Jarvik was particularly adept at the craft. He tested the Jarvik 5 on a series of cows, culminating with the successful case of Alfred Lord Tennyson, a calf who lived 268 days in 1981 with his artificial heart.

The public followed Barney Clark's progress with fascination, perhaps because he seemed to enjoy sharing the details of his life. Clark's father had died when he was twelve, and the boy took on a series of odd jobs—selling vegetables and hot dogs, and delivering newspapers—to help his mother keep their modest home in Provo, Utah. He met his wife, Una Loy, in the seventh grade. Failing to get into medical school at the University of Utah, he pursued dentistry in Seattle. A long-time smoker, Clark developed emphysema and hepatitis, then learned that his heart was failing. A physician attempting unsuccessfully to treat Clark's heart condition with drugs told him about the University of Utah's artificial heart program. Clark and Una Loy thought it over—quickly, for time was running out.

The sixty-one-year-old Clark read and signed an eleven-page informed-consent document twice, several hours apart, as the protocol required, indicating that he understood what would likely happen to him and what could go wrong. Years later, Corey would find himself reading and having explained to him the details of a forthcoming experiment, although his situation was not nearly as dire as that of the dentist.

If he wanted to live, Barney Clark really didn't have an alternative to signing the document and accepting the artificial heart. The heart surgeon who performed the procedure, Dr. William DeVries, told *Newsweek* magazine, "He was too old for a transplant, and there were no drugs that would help; the only thing that he could look forward to was dying." But the bioethicist George Annas wrote in *The Hastings Center Report* that Clark's informed consent was "incomplete, internally inconsistent, and confusing." Specifically, the plan did not account for Clark's becoming mentally incompetent or unable to communicate his wishes. For example, Annas wrote, the document stipulated that Clark would need to sign future consent forms if other procedures became necessary, but ignored the possibility that he might be unable to do so.

Unlike the tin man in *The Wizard of Oz,* who receives a new heart and immediately jumps up so fit he can bellow out a tune, Barney Clark had an agonizing final 112 days. During surgery, a lung leaked, his brain seized, a heart valve broke, and the anticlotting drugs triggered a nosebleed so severe it also required surgery. Afterward, Clark was unable to go home, as he had been told, and instead lay in the hospital bed pierced with hoses and tethered to a noisy 375-pound air compressor that enabled his new ticker to keep on ticking despite the unanticipated complications. Clark fought infection after infection. He had strokes as his blood platelets sheared against the uneven internal surfaces of the new heart, releasing powerful clotting factors. On March 21 pneumonia sent him to the intensive care unit, where soon his kidneys shut down as his temperature rose. Two days later, he went into multiorgan failure and died at 10 p.m., when his circulatory system collapsed.

Clark's comments both before and after the procedure suggest that, contrary to the bioethicist Annas's analysis, he considered his consent "properly informed." When asked how he would respond to other heart patients asking if they should have the procedure, he answered, "Well, I would tell them that it's worth it, if the alternative is they die." He also often expressed an altruistic spirit, telling a reporter from *Time* magazine just weeks before he died, "All in all, it has been a pleasure to be able to help people." Indeed, today more than a thousand people are alive because of left ventricular assist devices ("LVADs") that take over the functioning of the heart's hardest-working chamber—thanks to Barney Clark and others who tested prototypes.

Seven months after Barney Clark died, Stephanie Fae Beauclair was born, missing half her heart. Her physician, Leonard Bailey, had lost dozens of newborn patients to the condition, called hypoplastic left heart syndrome. Frustrated by the lack of treatments, he had begun experimenting with surgery seven years earlier, swapping hearts among goats, sheep, and baboons. In December 1983, an institutional review board (IRB) at Loma Linda University Medical Center evaluated the safety and ethics of his request to transplant a baboon's heart into a newborn. They said yes, because human hearts for such young infants were in very short supply.

By the time of the IRB approval, Baby Fae was two weeks old and close to death. She was on a heart-lung machine, with her temperature lowered to ease the burden on her body as she waited for someone else's tragedy to give her a heart. On October 26, 1984, while surgeons prepared the infant, a member of the transplant team went down to the medical center's basement and removed the apricot-sized heart from a young baboon, placed the organ in a dish filled with slushy, salty ice water, and raced back upstairs. Dr. Bailey gently lifted the baboon's heart and placed it into the baby's open and empty chest, then stitched its connections to the tiny blood vessels. When Baby Fae's body temperature was raised, her new heart began to beat.

A fascinated world watched the dark-haired infant with the baboon heart. Some people cringed at the idea of putting an organ from another species into a human, although thousands were already walking around with pig heart valves. Bioethicists, animal rights activists, and interested others came out of the woodwork to protest or proclaim, but the event ended quickly, for Baby Fae lived only twenty-one days with her new heart. Like Barney Clark, she succumbed not to immune rejection or the sheer strangeness of replacing a heart, but from an ancient enemy: infection.

Like in the case of Barney Clark, bioethicists and journalists questioned Baby Fae's informed-consent process. The science writer Claudia Wallis raised the issue in *Time* as to whether the research team had explained to Baby Fae's parents that there was a surgical alternative to a transplant to correct their daughter's heart defect. Ironically, the surgery was being done at Children's Hospital of Philadelphia, where several years later Corey Haas would be treated. Charles Krauthammer, also writing in *Time*, called the entire affair "an adventure in medical ethics." Krauthammer is a psychiatrist who was paralyzed in a diving accident, and writes often on medical ethics. Could parents of a desperately sick newborn, he asked, really think clearly enough to decide anything? This question would resurface frequently in the course of gene therapy, because most clinical trials—that is, experiments—target young and often desperately ill children.

The tales of Barney Clark and Baby Fae provide context to Jesse Gelsinger's death in a gene therapy experiment. The basic question is one of intent. *Is the experiment meant to help the particular patient or to advance a therapy that one day could help many?* Or both? This is a central tenet of medical research that is often misunderstood, especially by those outside the field.

Like Corey Haas, Barney Clark and Baby Fae happened to be in need at a time when an experimental therapy had proven itself in animals and was ready to be tested in humans. Corey, Clark, and Fae also had the good fortune to come to the attention of physicians who were plugged into the research going on in their fields. In the best of circumstances, the experimental subject understands the goal of the greater good, and that he or she may not in fact benefit from the treatment. In this way Barney Clark's bravery led to the left ventricular assist devices used today; baboon hearts never made it to the medical mainstream; and Corey Haas's restored vision will very likely lead to help for many. But this reality of medical research, the possible sacrifice of the few for the many, doesn't mean that the personal physicians involved in these complex cases didn't care profoundly about the patients and their families. They did.

5

JESSE AND JIM

JESSE GELSINGER WAS BORN ON JUNE 18, 1981. ARRIVING barely a year after his brother, Jesse didn't get the close attention of a firstborn, yet he hit all the milestones listed in the baby books: sitting, crawling, and babbling, which he quickly learned especially delighted onlookers, and walking right on schedule. The only unusual thing, his father, Paul, realized later, was the fact that Jesse was a very picky eater, sticking to potatoes and cereal and refusing meat and dairy.

Three months before he turned three, Jesse came down with a cold and his normally pleasant behavior suddenly changed. It was almost as if the smiling redheaded toddler had entered a time machine and emerged an angry adolescent. "His speech was very belligerent . . . as if possessed," recalls Paul. He and his wife, Pattie, took their son to the pediatrician. The doctor too quickly diagnosed anemia, sending the family home with instructions to feed the child milk, bacon, and peanut butter, the very things, his parents would learn later, that Jesse's body had been telling him to avoid.

In retrospect, it wasn't surprising that the push of protein would send Jesse into a coma. On a Saturday morning in mid-March, his parents found him curled up on the couch in front of the TV, sleeping so soundly that they couldn't wake him. Alarmed, Pattie insisted that they skip the local doctor and drive across the Delaware River to the children's hospital in Philadelphia. There, emergency department physicians found the boy responsive to stimuli but not awake—he was in a first-stage coma.

When initial blood tests indicated elevated ammonia, the doctors said it was Reye's syndrome, an illness recently recognized and linked to aspirin use. Ammonia in the blood was a sign of Reye's, but Paul and Pattie, with the sixth sense of parents, weren't convinced their son had Reye's, and additional blood tests proved them right. A week later, they had their answer: Jesse had an "inborn error of metabolism" called ornithine transcarbamylase deficiency (OTC) syndrome. The Gelsingers had never heard of it—only 1 in 40,000 people has the condition. Only three hundred patients had been diagnosed since the disease was first described in 1962.

Jesse's cells couldn't process dietary protein because his liver didn't make enough of an enzyme, OTC. Normally, OTC sends the nitrogens stripped off the amino acid building blocks of proteins out of the body as part of the urea in urine. Without enough OTC, liberated nitrogen combines with hydrogen to form NH_3—ammonia. Too much ammonia in the blood harms the delicate nervous tissue of the brain, and this is what had plunged Jesse into a coma.

Unlike Corey's disease, in which Nancy and Ethan each contributed a mutation, the mutant gene that causes OTC deficiency syndrome is on the X chromosome only and is transmitted from mother to son. Most moms aren't affected because as carriers they have a second X that provides enough of the enzyme. However, sons of carrier mothers who inherit the X with the glitch have the disease because they don't have the protection of a second X chromosome. Instead, they have the Y chromosome that makes them male.

Jesse, it turned out, had mild OTC deficiency, and it originated in him—he didn't inherit it from Pattie. Jesse was a "new mutation," and genetic testing also revealed that he was a mosaic—only some of his cells had the mutation. When Jesse was an embryo consisting of only a few cells, DNA in one of them spontaneously mutated. As the cell divisions of prenatal development continued, the mutation passed only to the descendants of that original errant cell. Jesse's partial disease was why he'd survived infancy. About half of children born with OTC deficiency die by the end of the first month, many in just a day or two as they lapse into comas. About half of those who survive infancy die by age five, even when they take drugs that stifle flare-ups. Few make it to adulthood.

Day-to-day life is challenging, both physically and psychologically, for a person with OTC deficiency. The child suffers frequent vomiting and stomach pain, requiring tube feeding when appetite is nil and meds must be kept down. If rising blood ammonia or a viral infection sends a child to the hospital, the parents wait in terror because they know each visit could be the last. At other times, children must follow a diet that one mother calls simply "incompatible with life." Among other restrictions, no hamburgers, no pizza, no hot dogs, and no ice cream.

During that first hospital visit, drugs brought Jesse's blood ammonia levels under control in a few days, and he went home on day eleven, with medications and strict dietary instructions. He stayed well until age ten, when a dietary lapse landed him back in the hospital for a few days. In 1987, the family moved to Tucson, Arizona, and changes followed: Paul and Pattie divorced, and Paul received custody of their four children. Paul married Mickie in 1992, adding her two children to his four.

Twice a year, Jesse attended a state-funded clinic for people with metabolic diseases. In September 1998, Jesse's senior year in high school, a specialist at the clinic, Randy Heidenreich, MD, mentioned a gene therapy trial for Jesse's disease about to get under way in Philadelphia. Participants had to be over eighteen. Jesse, still seventeen, seemed mildly interested. He was busy, with a part-time job in a supermarket and a motorcycle to care for, and he felt okay. Although he had to take sometimes fifty pills a day and watch his diet, he hadn't had an attack in a long time. Paul, however, was concerned that his son wasn't taking proper care of himself. Jesse wasn't running out of medications as often as he should have been, which Paul noticed because he was the one who picked up prescriptions. "Jesse was stressing his metabolism as he had never done before," he recalls.

Jesse's father's instincts were right. Three days before Christmas, Paul came home to find his son writhing on the couch and vomiting uncontrollably as a petrified friend looked on. They rushed Jesse to the hospital, where blood tests found his ammonia level to be six times normal. He was admitted to the hospital, and over the next few days he grew worse; at one point he even stopped breathing. With new medications, he finally came around, and he went home as 1999 began. It was to be Jesse's last New Year's.

The pricey new meds, just recently developed, lowered Jesse's blood ammonia levels and kept them manageable, and soon his appetite returned. He even weathered a bout of flu that February with no lasting effects other than giving it to his dad. Still shaken from the brush with death at the new year, father and son paid closer attention when Dr. Heidenreich brought up the gene therapy trial again at the next metabolic disorders clinic, and signed up for more information. In May, Dr. Heidenreich set up a visit for Jesse for late June in Philadelphia. He would finally be eighteen, and thus eligible to participate in the trial. It was exactly what would happen to Corey a few years later—a doctor would mention gene therapy once, and a family who'd never heard of it would wind up in a clinical trial after a second mention.

Fortunately, gene therapy is easy to understand. It is like replacing the corrupted part of faulty software or correcting a typo in an instructional manual. But practically, gene therapy is difficult to do. Researchers must know enough about a disease to figure out exactly which part of the DNA to replace, in which body parts to do it, and how to deliver the healing genes and get them to halt or reverse the disease process.

Early attempts at gene therapy delivered DNA to cells in tiny fatty bubbles or blasted it in with small gunlike contraptions. After much trial and error, most researchers settled on putting healthy human genes into viruses, then sending these viruses into patients' cells. Here, the Trojan-horse-like viruses send their genetic material into the nuclei of the cells, which is what viruses naturally do. A few types of viruses, including some that cause problems such as the common cold and AIDS, have been turned to the good side, delivering gene therapy.

We usually think of viruses as making us sick, but in gene therapy billions of them are sent into the body as microscopic ferries to make us well. The power of a virus is all the more astonishing considering its streamlined physique. Not a cell and technically not even alive, a virus is but a snippet of DNA or RNA wrapped in a thin protein shell, like a tiny Tootsie Roll Pop. As tiny as it is, a virus can wreak havoc on a human body if present in large enough numbers. That's what happened to Jesse Gelsinger.

On Saturday, June 18, Jesse's birthday, the family headed east to celebrate with some of Paul's fifteen siblings in New Jersey. On Tuesday they went to the Institute for Human Gene Therapy in Philadelphia and met with the surgeon Steve Raper, one of the three principal investigators—"PIs"—in the study. It took about forty-five minutes for Dr. Raper to go over the procedure and explain the informed consent document. Two catheters would enter and exit Jesse's liver, one to introduce the viruses bearing the healing genes into the hepatic artery, and the other to monitor whether the viruses—the "vectors"—were staying in the liver. Part of the rationale for the trial was to test the feasibility of sending genes into the liver. If it worked, the delivery mechanism might be useful for treating other urea cycle disorders, and for the many other conditions that affect any of the liver's vital functions, including blood clotting, cholesterol production, and handling toxins.

After the liver infusion, Jesse would have to lie still for about eight hours, and he would probably develop flulike symptoms as his immune system adjusted. Since the viruses would slowly leave Jesse's cells, these effects weren't expected to last long. Serious adverse effects, such as hepatitis or needing a liver transplant, were a remote possibility. To Paul, the worst part seemed to be the painful liver biopsy that would be done a week after the infusion to be sure that the viruses bearing their corrective cargo had actually remained in Jesse's liver cells. The clinical trial had been carefully set up so that every three patients treated received slightly higher doses, to figure out the least number of viruses that could be transferred safely. If symptom control improved, so much the better.

Paul, who had some science background, studied the details of the disease and the clinical trial protocol. "I said to Jesse that he needed to read and understand what he was getting into, that this was serious stuff," he recalls. Jesse knew that the therapy probably wouldn't help him, Paul maintains, but that he wanted to "help the babies" born with the syndrome in the future. Even if the effect faded within weeks, it might get infants past the early comas, buying time. Jesse's clinical trial would test the gene transfer—researchers didn't like to call it therapy until it clearly worked—on people with milder cases, such as women who were carriers and men like Jesse who had partial enzyme deficiencies. Jesse's altruism led some bioethicists to later question whether his consent had been truly informed.

After Dr. Raper explained the procedure, Jesse drank a small amount of ammonia tagged with a slightly heavy form of nitrogen, to see how much of it showed up in his blood and urine. This test would reveal how well his body handled ammonia, providing a baseline for comparison postprocedure. Past tests had shown Jesse's enzyme working at about 6 percent efficiency. While the family waited for the marked ammonia to appear in Jesse's body fluids, they toured the city, visiting the Betsy Ross House, the Liberty Bell, Independence Hall, and South Street. Mickie took a photograph of Jesse on the famous "*Rocky* steps" at the Philadelphia Museum of Art, named for their role in the 1976 film. The excited eighteen-year-old was trim, happy, and handsome, with his father's Kennedy-like good looks.

Once Jesse supplied the requested body fluids, the Gelsingers returned to Tucson and waited. The call came a month later. The second PI, Mark Batshaw, MD, delivered the news that Jesse had been accepted into the clinical trial. He would be the nineteenth patient enrolled.

Tall and slim with a shock of graying hair and a warm smile, Dr. Batshaw traces his fascination with urea cycle disorders to 1973, when he diagnosed a child with OTC deficiency syndrome whose symptoms had stymied others. He went on to pioneer drug treatments for it, including the new ones that Jesse was taking, and was the one who came up with the idea to try gene therapy. When Dr. Batshaw called the Gelsinger home that July day in 1999, he asked to speak to Jesse alone, since his patient was now over eighteen, but Jesse requested right away that Dr. Batshaw explain it all to his dad. That conversation, and what was misconstrued or not said, would turn out to be critical.

Dr. Batshaw confirmed that Jesse's enzyme was working at the expected 6 percent efficiency, which was the worst of all the people in the trial. The doctor explained that the gene therapy had worked in mice, that the most recently treated person had experienced a 50 percent increase in ammonia excretion, and if the gene therapy proved effective, twenty-five other liver disorders could be treated in the same way, amounting to some 12 million people worldwide. "Wow! This really works! So, with Jesse at 6 percent efficiency, you may be able to show exactly how well this works," Paul recalls saying to Dr. Batshaw, who agreed.

It was heady stuff. Paul doesn't remember any discussion of risks or dangers. Impressed by the numbers and overwhelmed with his son's desire to help others, Paul advised his son to say yes.

An early fall date was set for Jesse's infusion, and for the rest of that last summer, the new high school grad relaxed and enjoyed himself. A midsummer cover story in *Business Week* quoted a prominent scientist predicting that "within the next decade, there will be an exponential increase in the use of gene therapy."

It was not to be.

A trio of PIs led the clinical trial. Dr. Raper was the surgeon and Dr. Batshaw the metabolic disease expert. The third member of the team, who designed the gene transfer, would end up shouldering most of the blame for what went wrong with Jesse's treatment—James Wilson, MD/PhD, and today director of the Translational Research Laboratory at the University of Pennsylvania in Philadelphia, a facility that supplies viral vectors for gene therapy to researchers all over the world. Behind the doors next to the conference room where I spoke with Dr. Wilson, investigators continually look for new variants of the viruses that serve as gene therapy vectors, publishing their findings frequently in the journal *Human Gene Therapy*, of which Wilson is editor in chief, and elsewhere.

Jim Wilson is tall and lanky, with a broad grin that makes him look a little like the actor Jim Carrey. He grew up in Michigan, the son and grandson of physicians, and attended a small liberal arts college there in the mid-1970s. His interests in college soon turned from football to science—but the physical sciences, not the imprecise, descriptive biology. "I took one biology course and hated it," he recalls with a chuckle. But biology was on the precipice of great change, which Wilson realized during his senior year. Immersed in the first edition of the now-classic textbook *Principles of Biochemistry* by Albert Lehninger, Wilson saw that the new field of molecular biology was really biochemistry in a new light. "I became excited about the future of biomedical research. So I applied to graduate school for chemistry, and then, at the last minute, sent in applications to MD/PhD programs," he says.

Getting an MD/PhD at the University of Michigan won out over chemistry grad school. "I wanted to do good, fundamental biochemistry research in the context of human biology, which was the influence of my father and grandfather," Wilson recalls. The day he visited the department of biological chemistry to help with his decision, the chair, who never had a spare moment, gave him a precious hour. "Bill Kelley wrote the metabolic pathway for purines on the board, showing how defects in specific enzymes could lead to disease. BINGO." Wilson's career would be peppered with such moments of sudden insight when a syndrome on the whole-body level made sense at the cell or molecular level.

Wilson's memory of the seven years of oscillating between medical school and doctoral research are somewhat of a blur, and all he really recalls is that he worked in the lab all the time. He was investigating what went wrong with an enzyme called HPRT to cause a terrible disease called Lesch-Nyhan syndrome. Other researchers had approached the question using indirect tests, such as observing how antibodies respond to the abnormal enzyme, but Wilson wanted to know exactly how the sequence of the protein's amino acid building blocks differs in the abnormal enzyme compared with the normal. His mentor Kelley, although an expert on the disease, wasn't a protein person, and others in the lab told Wilson not to even attempt such a difficult project. It would take so long that he'd never get his PhD, they warned. This only prompted Wilson to find experts through Kelley's contacts, "in labs with awful smells, where organic chemistry was done by hand," to teach him how to snip the enzyme into pieces small enough to take apart and sequence.

Determining the amino acid sequence of the abnormal enzyme, using clues from the work of Theodore Friedmann, who'd isolated the gene, meant getting blood samples from patients. And that experience is what set Wilson's career path toward helping children.

"I got to know some kids well. It's just an awful disease. The children have a severe cerebral-palsy-like look, with abnormal movements. They have gout and often die of kidney failure, and are cognitively impaired. But the worst is the compulsion for self-mutilation. They bite their fingers off, and their lips. Parents tie their arms down because they try to bring their hands to their mouths, and have teeth extracted to prevent biting. They spit and swear, and have no control. Back then they'd end up in state institutions, where the caregivers thought that behavior modification would help," Wilson says, shaking his head in disbelief even decades later. Like Jesse's disease, Lesch-Nyhan is passed to sons from carrier mothers.

To find different mutations in the *HPRT* gene, Wilson traveled. "I spent a lot of time on airplanes, collecting cells from all over the world, going to clinics, drawing blood, processing the red cells, putting them on dry ice, and taking them in my luggage onto the plane and bringing them home."

A few of the kids would visit the lab, and everyone's favorite was fourteen-year-old Edwin. "He was quite a character. He lived in an institution, and he had a certain charm. At Michigan we took him to football games, onto the field. He was a king." One September when Edwin was visiting, Wilson finally figured out Edwin's exact mutation, and how it destroyed the key part of the HPRT enzyme. It was tremendously exciting because, at the time, very few mutations had been linked to their corresponding abnormal proteins. "And everyone had said we couldn't do it," Wilson says with a smile.

But Edwin's visits became tough on Wilson when he had to return the boy to the institution. "Edwin had a clean bed, food, and attention in Michigan, and then he'd have to go back. He became agitated, which is part of the illness. I'll never forget one plane ride back. He spat at the flight attendant. At home then there was no terminal, just a shack. Edwin was taken out on a gurney, down a steep set of stairs, to where his mother waited. I was so excited about finding the mutation that I blurted to his mom, 'I've got great news! We've made a great accomplishment! We know the mutation!' But she just looked at me, expressionless. And after a bit, she said, 'How is that going to help Edwin?' "

The moment was another turning point in Wilson's career, as he confronted the dual role of the scientist-physician. "I had this sinking feeling: how could I have been so selfish? All that work and money spent, and it would have no impact on the kids. I decided I've got to figure a way so every ounce of energy goes to directly helping people, not just for the sake of science. But I wasn't sure how I was going to do that." Recombinant DNA technology—using scissorlike enzymes to cut and paste DNA from one source into another—showed him how.

In 1983, Friedmann's group had sent functional *HPRT* genes into cells from a Lesch-Nyhan patient and coaxed the cells to make the enzyme. "This disease was the first to demonstrate the proof of principle that you could add a gene and correct a defect," Friedmann recalls, still excited at the memory. But the experiment corrected skin fibroblast cells, cultured in dishes because they readily divide—which is not the same as treating the nondividing brain cells of a child who chews off his hand or bites off his toes. Even though

Lesch-Nyhan led to the field of gene therapy, Friedmann says, "it's like an onion. The more we peel away, the more we want to cry because it's so complicated." The difficulty in translating encouraging results at the cellular level to making a disease's symptoms go away would be a repeating theme in the gene therapy story.

Wilson did his internship at Massachusetts General Hospital, where he supervised a young medical student named Jean Bennett, who would one day become Corey's doctor. Wilson had chosen Boston because it was an emerging center for gene therapy, and he went on to work in the lab of one of the leaders, Richard Mulligan, at MIT. It was here that his focus shifted to the liver with the headline-making case of six-year-old Stormie Jones, who received the first heart-liver transplant on Valentine's Day in 1984. She died six years later when her body rejected the new organs.

Stormie had the severest form of familial hypercholesterolemia (FH). Her liver cells lacked the receptors that admit the "bad" low-density lipoprotein (LDL) cholesterol, causing it to back up into her bloodstream and be deposited in telltale yellowish fat lumps behind her knees and elbows. Her high blood cholesterol, ten times normal, caused two heart attacks. Stormie's condition is extremely rare and caused by inheriting two mutant copies of a specific gene. Using FH as their model disease, Wilson and Mulligan were the first to deliver genes to hepatocytes—the most abundant type of cell in the liver. They even conducted a limited clinical trial to correct FH cells outside the body and put them back into five patients, who improved only slightly. Wilson continues to work on both Lesch-Nyhan disease and FH.

After his stint in Boston, Wilson returned to the University of Michigan, as faculty. He collaborated with another young researcher, Francis Collins, who was zeroing in on the gene, called *CFTR*, that causes cystic fibrosis, one of the most common single-gene diseases. When a speaker at the annual meeting of the CF Foundation in Orlando in 1989 canceled at the last minute, Collins sent Wilson to face the adoring crowd, just a week after the announcement of the CF gene discovery. "I walked into the meeting, and it was unbelievable. For better or worse this [discovery] was going to tell us how to cure the disease. It was almost a religious experience for me. I was just a young faculty member," Wilson says. CF was yet another disease that would prove challenging to treat with gene therapy because the corrections, to cells in the respiratory passageways, tended to last only a few weeks.

Through his encounters with Lesch-Nyhan disease, FH, and CF, Wilson was becoming an expert on the viruses used as vectors to introduce healing genes. The designation *vector* comes from traditional usage in referring to a carrier of a disease-causing organism, such as a mosquito carrying the parasite that causes malaria.

The handful of viral vectors used in gene therapy differ from one another in several ways: the types of cells they infect, whether they have DNA or RNA, their capacity, and whether they nestle into a human chromosome forever or just hang around in the host cell on their own, offering their cargo—the correct gene—for use, as the vector becomes diluted as the host cell divides. But perhaps the most important characteristic of a viral vector comes into play before it even enters a human cell—whether the surface hills and valleys form a pattern that provokes the host immune system or one that remains under its radar.

In 1990, Francis Collins inserted functional *CFTR* genes into retroviruses and then sent them into cells growing outside the body, or *ex vivo*, a first step to gene therapy. Typically the genetic information in cells and other types of viruses is part of DNA, which is copied into RNA, which directs synthesis of a specific protein. Retroviruses are so-called because they work backwards. Their genetic material is RNA, which host cells copy into DNA and then insert into their chromosomes. Collins's *ex vivo* treated cells indeed made the protein that people suffering from cystic fibrosis lack. By 1993, Dr. Ronald Crystal at the NIH had gone a step further, delivering working *CFTR* genes directly to patients' airways (*in vivo*), but using a different vector, adenovirus, or AV. This approach worked, but not for very long, because the immune system attacked the infected cells. AV is one of several types of viruses that cause the common cold and more serious respiratory infections, and although the virus had been stripped of its disease-causing properties, its membrane structure still provoked an immune response from the host. The patients' airways were responding to the sight of a Trojan horse, even though the horse had been emptied of its dangerous Greeks. Wilson was also working with AV, at high doses in monkeys, and although the virus effectively delivered its payload, he noted that it caused inflammation. In fact, the scientific literature had several reports of AV triggering inflammation in primates.

While Wilson was at Michigan, his mentor Bill Kelley moved to the University of Pennsylvania and immediately began recruiting Wilson. In March 1993, Wilson made the move, to head up the new Human Gene Therapy Initiative at Penn. There, he joined three key people who would greatly impact the history of the field: Jean Bennett, his former med student, who wanted to work on gene therapy for eye diseases; Guangping Gao, a postdoctoral researcher who found a way to tame adenoviruses; and Mark Batshaw, who'd been thinking about gene therapy for the urea cycle disorders since 1988. Wilson and Batshaw became excited at the idea of trying to treat a mouse that had partial OTC deficiency. Would healthy OTC genes carried in adenovirus work?

AV was well studied as a vector. It was large enough to hold a stowaway gene, yet would stay in a cell for only a few weeks before quietly disappearing, so any adverse effects—such as an immune response—wouldn't last long. Most important, the scientists thought, AV's surface features would lead it directly to the liver, where it could fix the hepatocytes that couldn't make the OTC enzyme. Mice given AV carrying OTC genes did very well, even eating high-protein chow without suffering any apparent digestive distress, for up to three months—a significant chunk of their two-year life span. But on the downside, a 1994 report in the journal *Immunity* described experiments in which mice treated with AV had a fast immune response—inflammation—that destroyed the targeted hepatocytes. The article concluded: "A better understanding of the response of the host to *in vivo* gene therapy is important in evaluating its usefulness in humans."

In 1994, Batshaw spoke about the mice that responded well to the delivered genes at the annual meeting of the National Urea Cycle Disorders Foundation. Mothers would be good experimental subjects, he told the families, because they carried the mutation and their responses could be measured as enzyme levels. That is, if the treatment worked, their enzyme levels, half that of normal yet not too low to cause symptoms, should rise. Grateful moms lined up to offer their blood, finally finding something they could do that might help their children. After the meeting, Batshaw and Wilson went on the road, speaking at medical conferences and rounding up more candidates by word of mouth in the pediatric community.

Having received overwhelming support, Batshaw and Wilson returned with renewed energy to their experiments with OTC-deficient mice. An enzyme deficiency is a little like a traffic jam in the body's biochemistry. The material that the enzyme normally breaks down instead builds up, and the product that should result from normal enzyme activity is absent or scant. In their pursuit of curing OTC, Batshaw and Wilson had developed drugs that lowered the levels of the excess metabolites (ammonium and glutamine) and boosted the deficient ones (citrulline and arginine), correcting a biochemical imbalance that saved Jesse Gelsinger in the middle of his senior year. But sticking to the daily complex drug-and-diet regimen was difficult, especially for children and teens. Batshaw and Wilson knew a gene therapy, a permanent solution, would make lives much easier, and with the correct gene working in mice, carried by AV, the goal of a forever fix in people seemed within reach.

The results in mice suggested that gene therapy for OTC deficiency syndrome could indeed work, because hiking the enzyme level just a little went a long way toward producing normal urine. If gene therapy could boost a 6 percent enzyme efficiency, such as

Jesse had, to perhaps 10 to 20 percent, then urea production might rise even higher. A kid might be able to eat a hot dog at a birthday party and not wind up in the hospital, and a small correction could perhaps even pull the babies out of their comas. Excited and encouraged, the researchers proposed the clinical trial in humans while continually refining the viral vector in mice, trimming the parts of the viruses that could infect the body or irk the immune system.

The OTC deficiency trial then had to navigate the regulatory maze. An Institutional Review Board (IRB) convened at the University of Pennsylvania approved the proposed clinical trial protocol in 1994. Senators Walter Mondale and Edward Kennedy had created IRBs as small groups of scientists and nonscientists in the late 1960s, to ensure that a proposed clinical trial complies with federal research regulations and adequately protects patients' rights. The Food and Drug Administration (FDA) and the Department of Health and Human Services monitor IRBs, and either agency can yank a trial. Biotech companies sponsoring clinical trials hire private review boards in place of IRBs. Today academic researchers increasingly use commercial boards to green-light clinical trial protocols, as the institutional groups become overloaded with projects to evaluate.

Once the OTC deficiency protocol passed the IRB, the next step was the NIH's Recombinant DNA Advisory Committee. The RAC had begun in the 1970s to review the safety of experiments that combine DNA from two sources, beginning with bacteria carrying human insulin genes to produce insulin for people with diabetes, and evolving to cover gene therapy trials. A RAC meeting is a public forum that convenes quarterly, and includes up to twenty-one unpaid experts drawn from genetics, medicine, and bioethics. Even if a proposed project is privately funded, researchers must submit information to the RAC if the institution where the trials will take place receives any funds at all from the NIH. That's almost everyone at a university-affiliated medical center. At a typical RAC meeting, several researchers briefly describe their projects, then field questions. Afterward, the RAC advises the other regulatory agencies—it doesn't issue an approval or denial of an experiment. In addition, investigators are obliged to report to the RAC a life-threatening adverse event or a death during an experiment within seven days.

Most of the members of the RAC that evaluated the OTC deficiency trial gave it high marks on several counts. The mutation and protein were well known, and symptoms stemmed from just one organ, the liver. Existing treatments were, even according to their inventors, "costly and unpleasant," and helped but didn't cure. Animal studies had shown that a small improvement in enzyme level could have a significant clinical effect. But there were some concerns. Two members of the committee asked about an experiment, eventually published in *Human Gene Therapy* with Wilson and Raper as coauthors, that showed that very high doses of adenovirus bearing a marker in place of a healing gene, administered to the liver of rhesus monkeys, caused "severe toxicity, including mortality." These types of studies laid the groundwork for adjusting the doses down for human use. Another concern was whether the viruses were safe enough to give to volunteers with only mild symptoms, such as Jesse. Why risk making someone sick who started the protocol in good health, with his disease under control? The RAC considered technical concerns, too. Was it better to deliver the loaded virus into the bloodstream or directly to the liver? The FDA finally ruled that the liver was best because it was the more localized route, but failed to inform the RAC of this decision.

Finally, in late October 1996, the FDA approved the trial, despite the concerns about the sick monkeys and the vector delivery route, and "invitations to participate" were sent to patients identified by their personal physicians. The trial was to be a "dose escalation" phase 1 pilot study to assess safety. Six groups of volunteers, each consisting of two women and one man, would receive the gene therapy with a pause of a few weeks between groups. If all seemed well in one group, the next would begin, with a fivefold increase in dosage. The escalation would continue until toxicity appeared, or symptoms or high ammonia levels abated, the best-case scenario. The first patient was treated on April 7, 1997. Jesse Gelsinger was in the highest-dose group, and would receive a seventeenfold-lower dose than the one that had harmed the monkeys.

6

TRAGEDY

THE DATE WAS SET, THE TRIP PLANNED. JESSE HAD originally been scheduled for the procedure in October, but a glitch with the patient before him bumped him up a slot, although he was still part of the trio slated to receive the highest dose.

Jesse had never traveled very far from home by himself, and he was excited as he packed new clothes and his favorite wrestling videos. He would fly out of Tucson on Thursday, September 9, and spend some time with his New Jersey cousins, in between blood and urine tests, before the infusion on Monday, the thirteenth. Paul would meet him in Philadelphia on the eighteenth, the day of the liver biopsy after the gene therapy, and they'd fly back together on the twenty-first. Paul wasn't frightened or uneasy when he dropped his son off at the airport. "As I walked him to his gate I gave him a big hug, and as I looked him in the eye, I told him he was my hero," he recalls.

Sunday night, September 12, the evening before the big event, Jesse's blood ammonia level was slightly elevated, requiring medication. When this had happened with other patients, the investigators had amended the clinical trial protocol to indicate that slightly elevated blood ammonia was "of uncertain significance." When Jesse related the elevation to his father, who was still in Arizona, Paul wasn't alarmed—at that point, he still trusted the medical team completely.

On Monday morning, a sedated Jesse was wheeled into the interventional radiology suite, where two catheters were inserted into his groin and maneuvered upward toward his liver. Then Dr. Raper administered 30 milliliters—about an ounce—of viral vector, from a different batch from the one used on the previous patient. The infusion took from 10:30 a.m. until 12:30 p.m. Afterward, Raper called Paul to say it looked like all had gone well. Paul and Jesse spoke briefly, each ending the conversation with "I love you."

Already, though, the viruses had done more than they were meant to. Deep in Jesse's liver, they were quietly alerting his immune-system cells that move among the more abundant hepatocytes, the intended target. The immune-system sentries, as researchers pieced together later from blood sampled throughout the two-hour infusion, included the gigantic blobby macrophages that wander the body seeking out trouble and dendritic cells that display molecular flags when invaders appear. "By the time we had withdrawn the catheter, there was evidence of systemic inflammation," says Wilson, with more than a decade of hindsight. But at the time, they didn't know this.

By Monday night, Jesse wasn't feeling well. He was nauseated and had a fever of 104.5 degrees. This wasn't unusual—all seventeen patients already treated had been fairly miserable the first night, with flulike symptoms.

At 6:15 the next morning, a nurse called Dr. Raper to report that Jesse seemed disoriented and his skin and the whites of his eyes had the yellow tinge of jaundice, a sign of a struggling liver. A blood test confirmed that Jesse's bilirubin—a breakdown product of old red blood cells, generated in the liver—was four times the normal level. The immediate danger was that the ruptured red cells could be flooding his circulation with globin proteins, which would break down, releasing ammonia. Raper called Batshaw, who was now at Children's Hospital Medical Center in Washington, D.C., and who hopped on a northbound train. Raper also called Paul. Had the team missed something in Jesse's medical history that would have indicated an additional liver problem? Had he ever had jaundice, a condition common among newborns? Paul couldn't remember, so he called Pattie, who did. Yes, their son had had neonatal jaundice. Meanwhile, Jesse's blood ammonia level was rising, and by midnight on that second day reached ten times normal. Jesse went on dialysis as his father got on a red-eye. "It's a very helpless feeling knowing your kid is in serious trouble a continent away," he recalls.

Jesse's condition worsened with astonishing speed as Paul flew east. By 8:30 Wednesday morning, Paul rushed to his son's bedside to find Jesse in a coma and on a ventilator. The two doctors coaxed the distraught father aside and gently explained that while the dialysis had lowered the ammonia level, other problems were arising. Tiny blood clots were forming spontaneously throughout Jesse's bloodstream, pulling clotting factors and platelets away from where they were needed, triggering bleeding in his vital organs and his skin. His breathing was in trouble, too.

By evening, the intensive treatment seemed to be turning things around. Jesse's blood and breathing were under control, so Batshaw returned to DC and Paul met one of his brothers for dinner.

Late Wednesday night, Jesse's blood oxygen crashed, and even with the supplemental oxygen jacked up as high as it could go, he couldn't draw enough into his lungs. By 1 a.m. Thursday, in growing desperation, Raper put Jesse on a heart-lung bypass machine or ECMO, for extra-corporeal membrane oxygenation. Because ECMO exchanges gases outside the body, more gently than the ventilator that forces air in and out, the idea was to give Jesse's lungs a rest. By this point, the chances of recovery weren't good. Left on his own, Jesse had about a 10 percent chance of survival; on ECMO, he'd have about a 50 percent chance. By three o'clock, though, he was clearly in crisis. Just as his lungs would recover a bit, his kidneys started to give out. By five o'clock, the ECMO was no longer working efficiently. The team struggled to keep enough vital organs going at one time to keep Jesse alive, while trying not to panic. On the other side of the hospital room door, Paul began frantically calling relatives—and a chaplain.

Mickie, Jesse's stepmother, was now on a red-eye, as Hurricane

Floyd swept up the East Coast from the Bahamas. Her plane was the last to land in Philadelphia before the storm hit, and she and the New Jersey relatives descended on the hospital. Batshaw was stranded on a storm-stalled train, desperately borrowing cell phones to stay in touch with Raper when his own ran out of battery power.

At noon, Thursday, September 16, Jesse's family was allowed into the room. "When we finally got to see Jesse, he was bloated beyond recognition," wrote Paul in a widely circulated document called "Jesse's Intent." "His eyes and ears were swelled shut, even wax extruding out of his ears. You know the clear fluid that comes from a wound? That happened in every cell in his body," Paul explained years later, after he'd become something of an expert in immunology. "His cells were leaking cytokines. They turned on and he couldn't shut them off." Jesse's body was drowning in a sea of his own inflammatory molecules.

Devastated, the relatives slowly left the room. Jesse's lungs were shutting down and his condition, according to staff, was "very grave."

Friday morning, Raper and Batshaw talked to Paul and Mickie about discontinuing life support; Jesse's brain had been irreversibly damaged. With the storm outside ebbing, a small group assembled in the hospital room. Paul, Mickie, and seven aunts and uncles drew near as Paul again called his son his hero. The chaplain said a final prayer. Ten staff, including the two doctors in charge of the clinical trial, stood at the back of the room, either crying or struggling to hold back the tears.

At 2:30 p.m., Friday, September 17, 1999, Dr. Raper stepped through the halo of relatives and turned off the machines, and they all watched the lines on the monitor fall and then flatten. The doctor who had placed the viruses into the young man's liver now placed his stethoscope on his patient's chest and murmured, "Goodbye, Jesse. We'll figure this out."

A little while later, Jim Wilson walked, head down, into Guangping Gao's office. "He was so sad. He felt very, very bad. He said, 'Guangping, we have to figure out what went wrong,'" recalls Gao.

And Gao, by now promoted to associate director of the gene therapy program at Penn, did just that—and his discoveries helped to make gene therapy safer. But they weren't in time to save Jesse.

Back in 1999, there were no texts or tweets, no Facebook to flash the news of Jesse Gelsinger's death everywhere instantaneously. Even if there had been, gene therapy was still an obscure corner of the biotechnology landscape to most people, if they'd heard of it at all. The news office at Penn didn't initially hold a press conference as it might today, most likely because everyone involved was in such a state of shock. With the typical delay in scientific publishing and exquisitely bad timing, the first relevant report was the protocol for the clinical trial, finally published in *Human Gene Therapy* with the date September 20, 1999: "Recombinant Adenovirus Gene Transfer in Adults with Partial Ornithine Transcarbamylase Deficiency."

Media coverage began on September 29 with "Teen Dies Undergoing Gene Therapy," the first of many articles in *The Washington Post*. It contained a few mistakes. Jesse did not mutate after birth, nor had he expected dramatic improvement in his health. Paul Gelsinger was called a "handyman," as if he fixes doorknobs and leaking sinks for a living, when in actuality he designs and constructs spectacular homes. Reporters called the physicians "gene therapists," as if the experiment was already a tried and true medical specialty, like proctology. The available treatments for OTC deficiency were depicted as simple fixes, not costly and difficult regimens that helped only temporarily.

Coverage of the story soon followed in the Associated Press, *The New York Times*, and of course the hometown *Philadelphia Inquirer*. The story detonated, and the headlines trace the trajectory from stunned grief to intensifying questioning:

PENN GENE THERAPY DEATH LEADS TO INQUIRIES,
The Philadelphia Inquirer, September 30, 1999

AFTER GENE THERAPY DEATH, INVESTIGATORS WONDER WHAT WENT WRONG, *The Scientist*, October 25, 1999

CALLS GROW FOR MORE OVERSIGHT OF GENE THERAPY, *The Washington Post*, November 24, 1999

Fear began to suffuse the field, impacting ongoing gene therapy trials. On October 11, the FDA ordered a "clinical hold," suspending some trials and blocking others from starting. Right away, the agency halted two trials sponsored by Schering-Plough, which were using adenovirus-based gene therapy for two types of cancer. Three patients had already suffered "serious adverse events," but the company had decided that these "SAEs" were "trade secrets," which the FDA is not required to release to the public. But in the wake of Jesse's death, the agency requested additional safety data from all clinical trials using AV, no excuses or exceptions, and discovered the Schering-Plough SAEs.

The journalists kept digging. On November 4, *The Washington Post* broke the news of seven other patients who had died in gene therapy trials, and the fact that their deaths had been reported to the FDA, but not to the NIH's Recombinant DNA Advisory Committee (RAC). Researchers countered that the deaths had been due to the diseases, not the treatments—a fact of life in many clinical trials for conventional drugs in which the patients are very sick. But the journalists seemed to hold gene transfer, perhaps because of its unfamiliarity, to a different standard.

Meanwhile, Paul and Mickie Gelsinger had quietly returned to Tucson. The Gelsingers had not, at first, blamed any of the researchers or physicians for what had happened. They'd seen the fear, frustration, desperation, horror, and grief mirrored in the medical team's faces. But the first ember of distrust ignited in early November.

Paul had invited Dr. Raper to spread Jesse's ashes on the young man's favorite mountain, planned for Sunday, November 7. The doctor flew in to lecture to Dr. Robert Erickson's genetics class at the University of Arizona on Friday. During the conversation after class, Paul started to get an inkling that there may have been warning signs. Erickson, it turned out, was one of the two members of the RAC for Jesse's clinical trial who had said that the experiment was not safe enough to proceed.

"The monkeys got really sick, Paul," Erickson said, and with that single sentence unleashed a seismic shift in the story.

"I hadn't heard any of this from Penn. That was my first indication that something was wrong. Then Raper said, 'We changed the vector to make it safer. That was our first generation that hurt the monkeys,'" Paul recalls. He had yet to learn that a second-generation adenovirus had actually *killed* a monkey. Ironically, Randy Heidenreich, Jesse's doctor at the metabolic clinic who had been trained by Batshaw and had first mentioned the clinical trial, had an office right next to Erickson's. Even though Heidenreich hadn't learned about the two RAC members who questioned the protocol's safety until after Jesse died, Paul still wonders how events might have turned out if Erickson had told Heidenreich about the monkeys, and Heidenreich—or anyone—had mentioned it when discussing the upcoming clinical trial with the Gelsingers.

On November 7, Paul, Mickie, various aunts and uncles, friends, and three doctors, twenty-five people in all, trekked up nearly ten thousand feet on Mount Wrightson, overlooking the desert with majestic evergreens in the distance. As Raper read a poem aloud, Paul released his son's ashes from a prescription bottle. "I will look to you here often, Jess," he said.

A few weeks later, Jim Wilson flew west, and he and Paul met for the first time. The clinical trial protocol forbade Wilson from meeting the patients during the trial because, as the inventor of the viral vector, he had a relationship with a biotech company called Genovo, which he helped found in 1992. This separation is a necessary barrier between academic experimental medicine and business. The Federal Technology Transfer Act of 1986 had set up CRADAs—Cooperative Research and Development Agreements—precisely so that a private company could help speed a research development toward clinical application. Wilson described himself, according to Paul, as an unpaid consultant to Genovo.

If Paul had been able to bring himself to read all of the news reports, he might have been better prepared for what would be revealed at the upcoming three-day RAC meeting to probe Jesse's death. Dueling headlines on December 2 glimpsed the looming conflict:

HOW A WORRIED MEDICAL TEAM PINPOINTED WHAT WENT WRONG, *The Philadelphia Inquirer*, December 2, 1999

RESEARCHERS CLAIM NO ERROR IN GENE THERAPY DEATH, *The Washington Post*, December 2, 1999

The newspapers summarized a voluminous report from the Institute for Human Gene Therapy at Penn that identified the cause of death as "an unusual and deadly immune-system response that led to multiple organ failure and death," with the most significant direct contributor an adult respiratory distress syndrome that hadn't been seen in animal studies. The report found no evidence of human error and concluded that the autopsy "revealed no information that would have predicted the events that led to Mr. Gelsinger's death." Still, the institute officially accepted responsibility. Two days earlier, the Center for Biologics Evaluation and Research at FDA had sent a letter to Wilson outlining the "specific violations" in the clinical trial. Batshaw and Raper received less harsh warning letters. The letter to Wilson began with a generalized "failure to fulfill the general responsibilities of investigators" and continued through a litany of all that had apparently gone wrong: inadequately informed consent, inconsistencies in the protocol, not testing sufficiently for elevated ammonia, and selective reporting of adverse events. The letter specifically spelled out the plight of "monkey AH4T," who developed body-wide, life-threatening tiny blood clots two days after receiving AV on October 27, 1998. Two other monkeys receiving slightly tweaked vectors had the same reaction. A final report from FDA on the matter states, "you failed to amend the informed consent document to inform prospective subjects of the possibility of this potentially life-threatening adverse event."

The RAC meeting on December 8, 9, and 10 was held at a conference center on the fringe of the sprawling Bethesda campus of the NIH and was open to the public. Paul was slated to speak on the final day, a Friday. On the days prior, he sat quietly toward the back of the packed auditorium, listening intently and scribbling on a pad of paper. Wednesday consisted of mostly technical comparisons of the varied viruses used in gene therapy. At that time, Paul still supported the researchers, at least one of whom was nearly in tears over a just-issued news release from the FDA blaming the researchers for the incompletely informed consent and a failure to report adverse events quickly enough, and questioning whether Jesse was an appropriate participant, given the hindsight of how quickly he became ill. Paul nevertheless told reporters, "These guys didn't do anything wrong." But his loyalty was soon to shatter.

The miscommunications and misunderstandings surrounding Jesse's death emerged a few at a time, on day two, as the technical details of sending billions or more altered viruses into the body were teased out. Some of it was confusing. For example, according to the protocol, Jesse received about 20 billion viruses—but that was per kilogram of body weight, not for his entire body. Jesse didn't weigh 1 kilogram—he weighed about 57. Paul, who hadn't realized how high the dose was at the time of signing the informed consent, now did the math.

Then there was the confusing matter of Jesse's slightly elevated blood ammonia level the night before the procedure. The level was in a nebulous zone of "normalcy," the definition of which had shifted back and forth during the various amendments to the clinical trial protocol. A more precise and consistently used definition of the normal range of blood ammonia level might have enabled the researchers to exclude Jesse from the trial beforehand.

Wilson explained that the point of an escalation trial is to see at exactly which dose range adverse effects begin to appear—and the people in the trial who had been treated before Jesse, including a woman who had received the same dose, had had only the expected flulike symptoms. But, apparently, people can react differently. The grisly outcome in Jesse's case completely surprised the team members. "At no time during or prior to this trial did we in any way expect to see what we saw in Jesse Gelsinger," Wilson told the crowd.

Juxtaposed headlines midway through the RAC meeting again highlighted the confrontational atmosphere:

FDA OFFICIALS FAULT PENN TEAM IN GENE THERAPY DEATH, *The New York Times*, December 9, 1999

PENN DENIES THERAPY LAPSE KILLED TEEN, *The Philadelphia Inquirer*, December 9, 1999

A third and very moving viewpoint at the RAC meeting came from an organization of families with urea cycle disorders. The two co-presidents, Cindy Lemons and Tish Simon, gave impassioned pleas to keep the gene therapy effort going, because the families had no other hope. They described life with kids who could never eat pizza or ice cream, and the constant fear that this would be the day that their children would not wake up. A seventeen-year-old with Jesse's disease spoke out. "Mom, please tell them I want to have gene therapy. I hate being sick."

As Paul listened, his uncertainty began to grow. Two issues finally pushed him over the edge. First was the fact that the sick and dead monkeys from other trials were not mentioned in the informed consent documents. The second alert came when Batshaw and Raper presented, on the last day of the RAC meeting, data from the clinical trial, and mentioned the woman who had shown a 50 percent increase in her ability to excrete ammonia. This was the fact that had convinced Paul and Jesse, in the phone conversation with Raper in July, to proceed—Paul thought it meant the therapy was working. But now, in December, he was hearing a quite different story. "At the meeting I found out that there had been no efficacy shown then. They said she had rectified on her own, and that it had nothing to do with gene transfer. That was the kicker, the straw that broke the camel's back," Paul recalls.

But that day, he chose to focus on those who had volunteered to have trillions of doctored viruses sent into their livers. And so with a strangled voice, to a hushed crowd, the father said, "All these people who participated in this trial did a wonderful thing. They came in with the same intent my son had. It doesn't get any purer."

The pieces of the puzzle had started to shift and assemble in Paul's mind and heart. On their own, the omissions in the information he and Jesse had received or the researchers' tone of overenthusiasm didn't seem so serious, but together they had added up to a tragedy. "I had been very supportive of these men and their institution after Jesse's death, but with this new knowledge, I was no longer able to support them," the distraught father eventually concluded.

On his way home from the RAC meeting, on the advice of one of his brothers, Paul called the New Jersey attorney Alan Milstein.

~

The new millennium brought apologies and sanctions. The FDA completed its investigation of the OTC deficiency syndrome clinical trial on January 19, citing eighteen specific violations, ranging from poor documentation of informed consent to eligibility forms filed after deadlines to the more serious inadequate adverse event reporting. Two days later, the agency temporarily shut down the Institute for Human Gene Therapy at Penn, placing eight clinical trials, including two for cystic fibrosis and *BRCA1* breast cancer, on clinical hold.

Alan Milstein had his first in-depth conversation with Paul Gelsinger on January 27. "What bothered Paul the most was that he thought they had lied to him about whether they had seen any efficacy prior to his son being in the experiment," the lawyer recalls.

Paul was also angry with Art Caplan, PhD, director of the Center for Bioethics at Penn and a member of the gene therapy team, for telling *The Philadelphia Inquirer* that the case would be good for teaching bioethics. The idea of reducing his son's sacrifice to an exercise in a textbook enraged the grieving father. But Caplan defends his comment. "I said that *scandals* tend to drive bioethics." (Unfortunately, Jesse Gelsinger has indeed joined Henrietta Lacks and Terry Schiavo as key case studies in many textbooks, but I dedicated the fourth edition of my genetics textbook to Jesse, in 2001, with his father's full approval.)

Paul Gelsinger also blames Caplan for steering the trial toward mildly affected individuals, like Jesse. "It would have been rational to test the babies because they had little recourse. They were going to die, and they could only benefit," Paul says. Many scientists agree with him. But Caplan's reasoning, which he stands by, touches on logic and logistics as much as ethics.

"The trial was never going to use dying babies, because the researchers wouldn't be able to see an adverse event. That is, if a baby died in the trial, no one would know if it was due to gene therapy or the disease. In order to figure out if a gene therapy vector has dangers, you have to try it on someone who isn't terminally ill," Caplan explains.

Another complication is that gene therapy isn't as easy to administer as other treatments—it's not a drug that can be injected or mixed into infant formula. A surgeon trained in the delivery technique would have to fly, on very short notice, to wherever a baby had just been born with this very rare disease, a luxury that government research grants don't support. "If we'd used babies, we couldn't plan because we couldn't predict when and where they would be born," Caplan adds.

According to the rules of pediatric research, Caplan says, for trials that carry more than minimal risk researchers should recruit people who are capable of consenting, and that means those with moderate or mild forms of the disease. The difficulty of explaining the trial to distraught new parents was paramount in Caplan's thinking. "Imagine a doctor has just told parents 'Your baby is going to die in the next twelve hours, we have a new thing, do you want to try it?' I was concerned that parents wouldn't be able to say no. It seemed inherently coercive to try to recruit participants under those conditions." There was no easy answer to who to recruit.

Perhaps the greatest source of misunderstanding, in Jesse's case in particular and in clinical trial design in general, is the intent of a phase 1 study: to assess safety. There is never a guarantee of a therapeutic effect. A trial participant might hope to benefit, but cannot expect to. Many patients, as well as journalists and even lawyers, do not understand this. In 2001, the National Bioethics Advisory Commission gave this common confusion a name: therapeutic misconception. It is "the belief that the purpose of a clinical trial is to benefit the individual patients rather than to gather data for the purpose of contributing to scientific knowledge." Caplan still questions whether Gelsinger's attorney truly understands the purpose of a phase 1 clinical trial. "Milstein believes you should never do research that is nonbeneficial. But that isn't what phase one studies are."

Efficacy is not a goal of a phase 1 trial, but identifying and promptly reporting adverse events is. Here the media were helpful. In the wake of Jesse's death, reporters found that of 691 adverse events in 93 gene therapy trials using adenovirus, only 39 had been reported to the FDA right away.

On February 2, 2000, a Senate subcommittee met for nearly four hours to go over Jesse's case. There, the clash between the search for a forever fix while "doing no harm" and biotech as a business was palpable. The former CEO of the biotech company Targeted Genetics, H. Stewart Parker, testified that the company bought Genovo, and Wilson received $13.5 million in stock for his 30 percent share, after Jesse died. Paul shared his perspective at the meeting too. He tearfully concluded that what happened to his son was "an avoidable tragedy from which I will never recover," although he is not against further research. Patients also implored legislators not to nix gene therapy. Eric Kast, a thirty-three-year-old with severe cystic fibrosis from Norman, Oklahoma, protested. "My battle with CF is a race.

Don't let me lose that race when the finish line might be just around the corner."

On February 17, the University of Pennsylvania issued a twenty-eight-page response to the FDA's findings of serious deficiencies, denying some of the minor infractions. Although an independent expert panel had suggested halting clinical trials for gene therapy at Penn completely, the center's effort was instead scaled back, and Wilson was banned from working on patients. By July some restrictions applied to his animal work as well, but he never stopped discovering new viruses.

A year and a day after Jesse died, Paul filed a civil lawsuit in Philadelphia against several individuals and entities: the three principal investigators (Wilson, Batshaw, and Raper); Penn and their Institutional Review Board that had approved the protocol; Genovo, Inc.; Children's National Medical Center; Art Caplan, the bioethicist, and William Kelley, the dean of the medical school who had recruited Jim Wilson to Penn.

On November 3, 2000, almost a year after Jesse's ashes were scattered over his beloved Arizona landscape, the lawsuit was settled, between the Gelsinger family and the University of Pennsylvania. Paul gave some of the money to the National Organization for Rare Disorders and the Citizens for Responsible Care and Research. Caplan and Kelley were released. The US Department of Justice also sued, settling on February 9, 2005. As part of that settlement, Jim Wilson had to write and publish an article reflecting on the experience, which he entitled "Lessons Learned."

Wilson crafted his article to address points raised in a scathing letter that he received from the FDA on February 8, 2002. The government document recapped the correspondence between the agency and the researchers, spelling out exactly what parts of the protocol were violated and why the team should have heeded warning signs. The highly personal letter was difficult to read, Wilson recalls, because Edwin, the boy with Lesch-Nyhan disease who had been just a few years younger than Jesse, was often still on his mind. Wilson says all he had ever wanted to do was to help sick children. In the FDA letter's six pages, the word *failed* appeared twenty times and *misled* nine times.

"Lessons Learned" finally appeared in the journal *Molecular Genetics and Metabolism* in February 2009. Although there was no deadline, the delay, Wilson says, was due to the legal matters, which didn't end completely until 2005. "Soon thereafter I started lecturing on the 'lessons learned' topic at many distinguished institutions, which I continue to do. These experiences allowed me to better formulate my thoughts." Also during those years, he continued research developing viruses to treat other diseases, including Corey's.

In the article, Wilson accepts some of the blame, but also deflects it. For example, the relevance of Jesse's elevated blood ammonia the night before the procedure, Wilson says, was a clinical judgment call. As for the monkeys, he, Batshaw, and Raper should have reported the toxicity to Paul, even though the problem hadn't arisen in an OTC deficiency trial but in one that used higher doses of a different viral variant. More philosophically, Wilson cited the team approach for minimizing individual accountability, and acknowledged that the researchers' overenthusiasm for their projects perhaps telegraphed undo optimism to desperate families seeking cures. Genovo, he explained, was a virtual company born of his doctoral work at the University of Michigan. He was not an employee nor on its scientific advisory board, and received no royalties, although he held the nonvoting stock. Targeted Genetics, which bought Genovo, ran into trouble during the ensuing dark years of gene therapy, paying nothing to its shareholders for some time before merging with a London-based biotech company in 2011, refocusing its goal on engineering viruses to combat antibiotic-resistant bacteria. But the fact that it offered shares to Jim Wilson at all suggested to Paul Gelsinger that the researcher was profiting from the experiment gone awry.

Aware of the appearance of financial impropriety, Wilson, in his paper, proposed a compromise: allow financial stakes for researchers involved in basic and preclinical work, but not for those directly involved in clinical trials. He also supported appointing third-party "patient advocates"—aka "child life specialists"—for the informed consent process, which many gene therapy clinical trials that enroll children use today.

Wilson looks back on the writing of the article as a true lesson learned, and not a punishment. "When you make a medical error, it's better to concede a mistake than not. Clearly mistakes were made and I felt responsible, but I wasn't allowed to talk. Having to write the article gave me time to reflect more on the responsibility we have as scientists and clinical investigators." Perhaps more important, though, than the details chronicled in "Lessons Learned," or even how bad everyone felt, was the fact that the field of gene therapy "was ahead of the basic understanding of the biological process. So we changed our focus to understand immunity and the host-vector interaction," says Wilson. To this day, we do not know all of the intricate steps in the choreography between a viral vector ferrying a healing gene and the immune system's sentry cells.

Jesse's father was less than thrilled with the punishment for those responsible for the clinical trial that led to his son's death: a five-year investigation by the Department of Justice translated into fines for the University of Pennsylvania and Children's National Medical Center of a little over a million dollars, no criminal charges, and temporary "restricted clinical activity" for the principal investigators (Wilson's professional activities were restricted for five years, and Raper's and Batshaw's for three). Paul has yet to see all the official records associated with his son's death, prompting him to write in *The Philadelphia Inquirer* on the tenth anniversary of Jesse's death, "Ten years ago today, my son died in a science experiment."

The sad case of Jesse Gelsinger has had two powerful and lingering effects. First, it led to improved protection of human subjects in gene therapy research. Changes instituted by March 2000 included new guidelines for conflicts of interest and informed consent, required bioethics training for researchers receiving NIH grants, mandatory redoing of informed consent for all participants if an adverse event occurs in even one, and submission of clinical trial monitoring plans to the FDA as well as the NIH. Second, Jesse's case radically changed the course of gene therapy, for it wasn't the healing genes that killed him, but the viruses used to deliver them. The concept of gene therapy prevailed, but attention would now turn to other viruses unlikely to provoke an immune response.

Paul Gelsinger still grieves, but harbors no ill will toward the goal of gene therapy. In fact, he draws comfort from knowing that his son's sacrifice has likely saved lives. "Jesse has made a lot of things possible." One of them is Corey Haas's newfound eyesight, courtesy of a much safer viral vector.

But Corey wasn't the first, nor the youngest, to undergo successful gene therapy. That actually happened for the first time eighteen years *before* Corey's surgery.

PART III

EVOLUTION OF AN IDEA

If you have people dying of genetic disease, due to a defective gene, then you correct the gene. . . . I am delighted that today, gene therapy is having a rebirth.

—W. French Anderson

7

THE SCID KIDS

CURIOUSLY, DATES IN MID-SEPTEMBER MARK THE milestones in gene therapy. Corey's procedure took place on the twenty-fifth in 2008, and Jesse Gelsinger's on the thirteenth in 1999. The very first gene therapy for an inherited disease happened on September 14, 1990. On that Monday afternoon, four-year-old Ashanthi DeSilva received an infusion of her own T cells, bolstered with working copies of a gene that her failing immune system desperately needed.

Ashi had her first infection at just two days of age. By the time she was walking, she was constantly hacking and dripping with coughs and colds, and just toddling would make her as winded as an elderly chain smoker, recalls her father, Raj. Doctors at first offered the common explanations: asthma, allergy, bronchitis. But standard treatments didn't work, and Ashi was still ill nearly all the time. Raj's brother, an immunologist, advised taking a closer look at her immune system, and that led to her diagnosis, just past her second birthday. Like Corey, Ashi was a medical zebra. She had ADA (adenosine deaminase) deficiency. Lack of the enzyme adenosine deaminase stopped a key biochemical reaction in a way that caused the buildup of a toxin, poisoning her T cells, the white blood cells that serve as linchpins of the immune response. These are the same cells that are the initial targets of HIV.

Also like Corey, Ashi was in the right place at the right time—twice. Shortly after her diagnosis, she started a new enzyme replacement therapy that worked well for two years. It wasn't a forever fix, but it gave her body the enzyme it couldn't make. As her T cell count approached normal, she gained weight and had longer healthy periods between infections. But gradually her T cell count fell, infections returned, and tests showed a waning immunity. Then the little girl got lucky again: she was selected to receive the very first gene therapy at the National Institutes of Health (NIH) in Bethesda, Maryland. In case it didn't work, though, her doctors continued giving her the enzyme replacement therapy.

Ashi is healthy today and need no longer be treated. Her ADA deficiency is a form of severe combined immune deficiency, or SCID. The SCID disorders have played a prominent role in the gestation, birth, and sometimes rocky development of gene therapy. The "SCID kids" were the first to have gene therapy, providing some of the groundwork for Corey's successful treatment.

The eight different types of SCID affect only 1 in 100,000 newborns and are classified by the types of immune system cells that they cripple. Understanding how a rare *inherited* immune deficiency kills T cells can provide insight into a much more common *acquired* condition that does much the same—AIDS. Only twenty children are born with ADA deficiency worldwide each year, whereas many millions of people acquire HIV infection.

When researchers began distinguishing the types of inherited immune deficiency in the 1960s and 1970s, they quickly saw that boys outnumbered girls. That meant that at least one form of the condition must be passed from carrier mothers to affected sons, like OTC deficiency, Jesse's disease. One boy with an X-linked form of SCID showed the world, long before AIDS, how terrible life is without immunity.

David Phillip Vetter, the "Bubble Boy," was born in 1971. It wasn't until 1993, however, that researchers traced his form of SCID, called SCID-X1, to a mutation in a gene that normally tells immature white blood cells in the bone marrow to make a type of protein that lies on their surfaces like catcher's mitts. The proteins "catch" molecules that signal the cell to mature and respond to infection by cranking out cytokines, the immune system molecules that, in crazy excess, killed Jesse Gelsinger. Before that discovery, researchers knew only that the rare disease passed from carrier mothers to sons, with a probability of 1 in 2. Boys with SCID-X1 never lived more than two years.

Unlike Corey's impaired vision or Ashi's constant infections, David's illness was not a surprise. His parents, David J. Vetter, Jr., and Carol Ann Vetter, of Shenandoah, Texas, had a healthy daughter, Katherine, in 1968, and a son, David Joseph Vetter III, in 1970. Their first son was always ill, and when doctors diagnosed him with SCID, they tested Katherine to see if she could donate bone marrow. She was indeed a match, but her brother died before her marrow could replace his failing cells. He was seven months old.

When Carol Ann became pregnant again, she knew a boy would have a 50 percent risk of inheriting SCID. When amniocentesis indicated that the fetus was male, the couple elected to take their chances and continue the pregnancy. They'd hope for the best but prepare for the worst. Their doctors planned to have sterile conditions in the delivery room and nursery until tests could reveal whether the baby had inherited the mutation. If he had, perhaps the child could participate in an experiment. Raphael Wilson, a biologist at Baylor College of Medicine, which was affiliated with Texas Children's Hospital, was ready to duplicate a bone marrow transplant experiment he had done on twins in Europe, who had developed immunity after three years in isolation thanks to the transplants.

On September 21, 1971, David Phillip Vetter was born by cesarean section at St. Luke's Episcopal Hospital in Houston. The three-room delivery suite had been scrubbed for days, and all who entered it were first tested for lurking germs. The delivery took fifteen minutes, proceeding in silence as the team communicated with limited movements to avoid disturbing the air and possibly flicking a stray bacterium onto the newborn. Nurses suctioned the baby and whisked him into a plastic isolator. Then he was carefully moved to a setup at the attached Texas Children's Hospital that had a "crib bubble" for David and a "supply bubble" for his paraphernalia, which had to be sterilized in a series of meticulous steps. Rubber gloves poked into the bubble, enabling nurses to reach in to care for the newborn.

The blood test results brought the worst possible news: David had SCID, and his sister's bone marrow wasn't a close enough match to help him.

David lived in a series of bubbles, at first mostly on the third floor of Texas Children's Hospital, then more at home, using a transport bubble in between. Friends and neighbors visited often. Katherine was very protective of her brother. She slept next to the bubble, and shielded him from the eyes of onlookers en route to the hospital. They even squabbled like normal siblings, with Katherine deflating parts of her brother's realm when she was mad at him, and David getting in a few deft punches using the attached gloves.

Psychologists and psychiatrists visited David regularly. The boy bonded emotionally with hospital staff, especially Raphael Wilson. When David was four and Wilson was away for several days, the distraught boy punched holes in the bubble. When Wilson missed time due to a heart attack, David, feeling abandoned, painted the interior walls of the bubble with his excrement.

David established what was to be a lifelong relationship with a PhD student, Mary Ada Murphy. She first met the three-year-old David at his home. Murphy's doctoral thesis was about stress in families with birth defects, and her mentor had asked her to accompany him to the house to administer intelligence tests to David. She sensed how to communicate with the child, and quickly earned David's trust. She tried to teach him what was outside his bubble world. When David referred to a tree as a green shape and a brown shape, she went outside and brought him a branch, to give him a sense of the magnitude and majesty of a real tree.

Murphy would work on her dissertation in David's room at the hospital. She became the boy's champion and confidante, privy to his deepest fears and aware that he was not at all the happy boy in the bubble the media portrayed. By age five, he knew he was different. At seven, he asked her why he was angry all the time, and spoke of the futility of everything he was made to do, like school lessons. He grew increasingly resentful of the medical staff, who seemed to expect him to be compliant and upbeat. In frustration, David asked Murphy to write his true story. She began taking notes.

By the time David was eight, his anger and frustration were dominating his life. Why was he ever put in the bubble? Why hadn't he any say in the matter? When David was nine, the medical team, which by this time consisted of different members from the original team, began debating what to do. Should David leave the bubble, and take antibiotics and gamma globulin to fight infections for as long as possible? Murphy supported the idea. David's parents did not, and they, with the original doctors on their side, won out. But things changed in 1983.

New drugs and ways to manipulate bone marrow had made it possible for David's body to accept his sister's imperfectly matched bone marrow after all. So on October 21, 1983, the precious pink fluid from his sister dripped into David's veins. He did well until December, when he fell ill, and in early February, he suddenly developed diarrhea, vomiting, and a high fever. His condition became so dire that he had to be taken out of the bubble. He left it on February 7, 1984, and was able for the first time to touch his parents and sister. When awake, he watched TV coverage of his own life and implored Murphy again to tell the true story. But he was growing more fatigued and seemed to know the end was near. David Phillip Vetter died on February 22, 1984.

The autopsy held a surprise: David had died from widespread cancer of the lymphatic system, lymphoma. He'd gotten it from Epstein-Barr virus in his sister's bone marrow, which at the time had been linked only to mononucleosis. The virus had never made her sick because her healthy immune system kept it at bay. David had no such defense.

~

The bone marrow transplant might have saved David Vetter by replacing his sick *cells*. *Gene* therapy offers a more precise correction, and by the time of David's death, Theodore Friedmann had already shown that it was possible, at least in cells growing in a lab dish. Several lines of evidence were beginning to converge to show that gene therapy might be a valuable treatment approach for any inherited disease whose mutation was well understood. Shortly after David lost his battle in 1984, researchers published the first human genetic marker map. By today's standards the map was a crude collection of signposts, but it nonetheless served as a starting point and scaffold for the human genome project that would get under way a few years later. Meanwhile, starting in 1989, a stream of discoveries of disease-causing genes ensued: cystic fibrosis, Duchenne muscular dystrophy, Huntington's disease, and more. On other fronts, researchers were dissecting the structures and functions of the viruses that could shuttle genes, and fine-tuning the technology of combining DNA from different sources. These were the tools of the fledgling field of gene therapy circa 1990, when Ashi had her treatment at the NIH.

In retrospect, replacing a defective gene to cure a disease seems obvious. But the idea took time to grow out of an understanding of the nature of the genetic material, and what it does. To those of us who majored in biology, the following few paragraphs will be as familiar as the ABCs. For a different perspective, read this section as the parent of a child just diagnosed with a genetic disease might, who has to suddenly master the basics of DNA science against a backdrop of shock and fear.

Our quest to understand the chemistry behind heredity dates to the first part of the last century, and culminated at the turn of this century with the sequencing of the first human genome. Midway in the twentieth century, James Watson and Francis Crick assembled clues, mostly from experiments done by others, and deciphered the structure of DNA. The molecule that passes traits from generation to generation, DNA—deoxyribonucleic acid—resides in the cell's nucleus, the centrally located membrane-bounded genetic headquarters. A DNA molecule consists of two head-to-tail sequences of four "letters," which are nitrogen-containing chemicals called bases. The two halves come together like a long ladder elegantly twisting to form a double helix, with the bases pairing to form the rungs and side rails built of alternating sugars and phosphorus-containing phosphate groups. The DNA is highly wound around proteins that serve as spools, so that if the DNA in the nucleus of a human cell were stretched out, it would be about as tall as an average man.

The four-letter chemical alphabet and the double helix structure immediately suggested how the molecule copies itself. Like a line dance, the long, symmetrical molecule parts down the middle, and as each piece of information brings in a new partner from bases floating freely in the cell, two double helices unfurl from one. The base sequence of a DNA molecule imparts information as a code that the cell uses to construct proteins, in three-letter words using a four-letter alphabet. A protein is a large molecule built of a string of amino acids that folds, forming a highly specific three-dimensional shape.

The four nitrogen-containing bases that are the building blocks of DNA—A (adenine), C (cytosine), G (guanine), and T (thymine)—can form sixty-four different groups of three, or triplets. Building a specific protein begins when a cell copies a gene's worth of DNA—typically a few thousand bases—into an intermediate form called messenger RNA, or mRNA. (RNA has the same bases as DNA, except it has U [uracil] instead of T.) The mRNAs exit the nucleus through pores, entering the jellylike cytoplasm and anchoring onto small balls called ribosomes. Here, cloverleaf-shaped connector molecules called transfer RNAs (tRNAs) swim up to the mRNAs, docking at certain ones by means of a three-base sequence, called an anticodon, that aligns base-by-base with the mRNA's triplets, called codons. All of this attaching happens because of chemical attractions called the base pairing rules: A binds U, and G binds C.

Genetic information "flows" from DNA (gene) to RNA to protein, and it is the protein that imparts the trait—such as a blood clotting factor. The key to how a gene encodes a protein is that each tRNA carries, opposite its anticodon attached to the stalk part of the clover-leaf, an amino acid. A particular tRNA always carries the same type of amino acid—there are twenty. As tRNAs link to an mRNA, the amino acids align and join, like children holding hands, creating a polypeptide. The choreography is a little like a strip of Velcro with pompoms (the tRNAs and their amino acid cargo) attached, that stick to another strip of Velcro (the mRNA that holds the gene's message). Some proteins, such as insulin, are just one polypeptide. Others, such as the hemoglobin that carries oxygen in the blood, are composed of several polypeptides.

The often-misused term *genetic code* refers to which DNA/RNA triplet specifies which amino acid. The code is universal—the same for every species. For example, CCG spells the amino acid proline in an elephant or an eggplant or even a virus. In gene therapy, it's crucial that the genetic code is also the same for the viruses, so that they can deliver healing genes into cells that can read them. Even viruses with RNA as their genetic material—the retroviruses that are so important to gene therapy—make enzymes that compel cells to copy viral RNA into DNA.

DNA, RNA, and proteins can all serve as drug targets, but only intervention at the DNA level is a forever fix. Several very promising new drugs to treat cystic fibrosis (CF) illustrate the difference between targeting the protein versus replacing its gene.

The clogged lungs and pancreas that are the hallmarks of CF are the result of misfolded or mangled proteins with the unwieldy name "cystic fibrosis transmembrane regulator," or CFTR. In the disease, the CFTR protein can't reach or work properly at the surfaces of cells lining the lungs and pancreas. The proteins are supposed to form channels for chloride, and when they are impaired, salt balance is off. Secretions become concentrated into a sticky muckiness that is enticing to microorganisms that don't ordinarily colonize human lungs, causing severe and difficult-to-treat infections.

The new CF drugs either correct misfolded CFTR proteins or escort functional proteins that are just stuck on the production line to the cell membrane, where they can work. Patients take two tablets a day and breathing eases. One manufacturer calls the drugs "new medicines to treat the underlying cause of cystic fibrosis," a phrase that trickled into the media as, for example, Forbes.com's "the first drug to treat the disease at its genetic root." This description is incorrect: the drugs correct the *protein*, twice daily, and not the *gene*, the misguided instructions that create the problem. The new medicines are a little like using Wite-Out to correct an error in many copies of a printed page. Gene therapy would correct the error in the template, the original document from which the copies are generated.

As anyone who has owned a car or computer can attest, being able to name the parts of something and being able to fix it are two very different skills. So it was with early gene therapy. The concept was straightforward, the execution not. In the 1950s and 1960s, many of the researchers who discovered the workings of the genetic material used bacteria, which have a single molecule of DNA unencumbered in a nucleus. But few people dared to dream about changing the genes of humans. One of them was William French Anderson.

Born in Tulsa, Oklahoma, Anderson preferred reading science books to almost anything else, advancing to college textbooks by the third grade. He was shy and stuttered. In the fifth grade, when a classmate walked home with him one day and blurted out that French was the most unpopular boy in the school, the comment shocked him into turning his life around. Instead of considering others too unintelligent to bother with, as he had done for years, he started to listen to them and gradually made friends. By seventh grade, he was elected class president, he excelled at track and field, and he joined the debating team.

Anderson traces his first thoughts on gene therapy to 1954. He was seventeen and on the brink of his first major accomplishment:

getting into Harvard. Watson and Crick had just solved the structure of DNA, and that told Anderson where his future lay—understanding inherited disease at the molecular level. By the time he hit the Harvard campus, he had planned his career as a medical researcher, in detail. Attired in cowboy boots and speaking in an Okie accent, he stood out on the staid Ivy League campus.

The winter of 1958 was the era of Sputnik, the Russian satellite that launched the space race. Science was suddenly cool. Although still an undergraduate, Anderson was allowed to participate in "journal club" for biochemistry. Journal club is a rite of passage for scientists, a weekly meeting of the hierarchy of a research group—the professors, postdoctoral research associates, graduate students, and a select few undergrads. Journal club gave Anderson practice in selling a strange idea. He had no inkling at the time how important that skill would be.

At journal club one day in 1958, a visiting postdoc was talking about the structure of hemoglobin, the four-part molecule that carries oxygen in the blood. Anderson couldn't contain himself.

"What if we found out what is wrong in sickle cell anemia? We could put in a normal globin gene and cure it!" he said excitedly.

The others turned and gave him withering looks. "This is a serious scientific discussion," piped up one of Anderson's superiors. "If you want to daydream, keep it to yourself!"

Anderson was devastated. But then he heard another comment. "Interesting idea!"

Despite the original deflating comment, Anderson wrote his senior thesis on what would become his life's goal, correcting genetic errors. He went on to work with the giants of the new field—Francis Crick at Cambridge University in the UK, and Marshall Nirenberg, head of the group deciphering the genetic code at the NIH. It was from Crick that Anderson learned a lesson that would remain with him for life: if a theory is sound, but the evidence doesn't support it, keep doing experiments. Find more evidence. It was in Cambridge that Anderson met his wife-to-be, Kathy, when the two shared a cadaver head in anatomy lab. She was the better dissector and went on to become a noted surgeon. They made a striking pair: she looked like the film star and princess Grace Kelly; he had the tall frame and good looks of the actor Jimmy Stewart.

At this time, at the height of the hippie movement of the 1960s, several people were thinking, speaking, and writing about gene therapy. Joshua Lederberg, who was awarded the Nobel Prize at age thirty-three in 1958 for discovering how bacteria swap genes, discussed the steps of gene therapy at Columbia University, then wrote it up in *The Washington Post*. Nirenberg penned a piece in *Science*, "Will Society Be Prepared?" that predicted gene therapy by 1992—not far off from what Anderson would accomplish. But others cautioned that the term *gene therapy* was premature. What was being discussed was *gene transfer*, with a goal of gene therapy.

The first attempt at gene therapy occurred in 1970, when Stanfield Rogers of Oak Ridge National Laboratory, with physicians in West Germany, used a virus that causes warts in rabbits to treat two girls with a rare inherited condition that causes intellectual disability, spasticity, and seizures. He hoped that an enzyme from the virus would replace the one that the girls were missing. Although Rogers's approach didn't work, the idea of using nonhuman genes to correct a disease in humans, and from a source that could make people sick in other ways, made some researchers uneasy. Theodore Friedmann, who would go on to demonstrate gene transfer for Lesch-Nyhan syndrome and impress a young Jim Wilson, called for further thought. In a 1972 article in *Science* he cautioned against attempting gene therapy too soon, before we knew the details of how genes work.

While the initial attempt at gene therapy in 1970 delivered a gene from a rabbit virus into two girls, the idea of using viruses to deliver *human* genes was building a following. It was an example of a new approach, called recombinant DNA technology, which combines DNA from different species. The idea set off so many alarm bells that it led to a famous meeting on California's Monterey Peninsula in February 1975. The 140 attendees, mostly molecular biologists, with a few lawyers and physicians sprinkled in, met at the Asilomar Conference on Recombinant DNA to discuss a new breed of experiment. Paul Berg of Stanford University wanted to attach DNA from a monkey virus called SV40 to DNA from a different virus that normally infects bacteria, and send it into *E. coli* bacteria. The potential danger was that SV40 causes cancer in mice, and *E. coli* inhabits human intestines. If altered bacteria settled into a human gut, might they churn out the virus, causing cancer? After heated discussion, the conferees concluded that the experiment was safe, but they established containment measures that persist to this day. Starting with human insulin manufactured in bacteria, recombinant DNA technology has since contributed a few dozen drugs and has never led to what one researcher dubbed "triple-headed purple monsters." The researchers who pioneered the technology couldn't have known how valuable the ability to mass-produce human proteins in bacterial factories would become once HIV showed up in the blood supply in the early 1980s. Furthermore, the Asilomar conference set the precedent for the biotechnology community policing itself.

Recombinant DNA technology began on bacteria, which are single-celled. Adding genes to many-celled organisms, which is more of a challenge, produces a transgenic organism. A broader term for life-forms with altered DNA is "genetically modified organisms," or GMOs. Corn that makes its own insecticide thanks to a bacterial gene, goats that manufacture a human clotting factor in their milk, and the many mouse "models" of human genetic diseases are GMOs. Transgenic organisms have a foreign gene in each of their cells, because they are manipulated as fertilized eggs. Other biotechnologies "knock out" a gene to see what happens when it's gone, or "knock it down" to diminish its effects.

Gene therapy in people has its roots in recombinant DNA technology too, but it doesn't create a transgenic person (although theoretically it could). Instead, human gene therapy alters specific types of body cells, after birth. In genetics lingo, Corey's type of gene therapy is *somatic* ("body"), which means that it won't affect future generations. The other type of gene therapy is *germline*, which affects sperm, egg, or fertilized egg, and is passed to offspring. For example, a cat's egg given a jellyfish's gene for green fluorescent protein, when fertilized, leads to a cat that glows greenly—as do its offspring. Corey received new genes into cells in his retina—not in his sperm-to-be.

It wasn't until the mid-1980s that researchers began figuring out ways to send genes into cells. Gene-laden fatty bubbles called liposomes could gently coalesce into cell membranes, like small soap bubbles joining a larger one, depositing their cargo. A tiny gunlike device could blast DNA into cells. Naturally occurring circles of DNA called plasmids could be opened up and genes of interest inserted, like children holding hands in a circle letting other children join. Then the loaded plasmid is delivered into a human cell's nucleus. The idea to add a human gene to a virus and then infect human cells came from Richard Mulligan at MIT, in 1984. Viral delivery slowly came to dominate, which was why Jim Wilson came to work with him.

The first attempt at gene transfer used a chemical to carry DNA into cells. In 1980, Martin Cline from the University of California, Los Angeles, wanted to try gene therapy for beta-thalassemia, a form of hereditary anemia common among those of Mediterranean descent. But when UCLA wouldn't approve the protocol, he did the experiment in Israel and Italy. In response, the NIH stopped his funding and he left UCLA. The treated patients didn't get better, but they didn't get worse. When the *Los Angeles Times* reported the unauthorized experiment, *The New England Journal of Medicine* hurriedly published an article that French Anderson and the bioethicist John Fletcher had submitted months earlier, since it was suddenly timely. Anderson and Fletcher urged that animal studies demonstrate that a gene gets into its target and stays put, not causing harm, before trying the approach on people.

In between stints at Cambridge and the NIH, Anderson returned to Harvard to earn his MD, where he continued to talk constantly about gene therapy, and where he became intensely interested in families with childhood inherited diseases. While in Nirenberg's lab at the NIH from 1965 to 1968, Anderson got to know a young brother and sister who had beta-thalassemia. Diagnosed as infants, Nick and Judy Lambis suffered from extreme fatigue. Their bones hurt from marrow overgrowing to keep up with the diminishing red blood cell supply. In 1968, Anderson started his own lab group at the NIH and took over the siblings' care. He wanted to apply his idea of gene therapy to save his young patients, but there wasn't yet a way to deliver genes. Instead, he helped invent the chelation therapy that thalassemia patients use to combat the iron buildup from many blood transfusions. Neither Nick nor Judy survived adolescence, and gene therapy for beta-thalassemia wouldn't work until 2010.

Beta-thalassemia was an early target for gene therapy because researchers already knew a lot about hemoglobin and anemia. Anderson had been desperate to use gene transfer to treat Nick and Judy Lambis and felt scooped by Cline's work, but both experiences made him realize that gene therapy, at least for beta-thalassemia, was not nearly as simple as it seemed. The challenge was that a hemoglobin molecule has four parts, two alpha and two beta chains. One gene specifies the alpha chains and another the beta chains. Beta-thalassemia affects the beta chains, but successful gene therapy would have to maintain approximately equal numbers of alpha and beta, and no one knew how to do that. In addition, cells bearing corrected genes would have to replace cells stuck with the mutant genes fast enough to have a noticeable effect. It was a tall order.

To test whether gene transfer could really provide gene therapy, researchers needed a simpler sickness. That meant a disease caused by a mutation that acts like an on/off switch, with an effect easy to see or measure. Most important, corrected cells should have an advantage over sick ones, so that they would not only persist, but take over.

ADA deficiency, Ashi's disease, fit the bill. It's a straightforward enzyme deficiency, affecting only one type of cell. It leaves something in the body that's detectable—uric acid, the stuff of bird droppings. The uric acid overload destroys T cells, which in turn cannot activate B cells to make antibodies. Immunity crashes.

A treatment for ADA deficiency—enzyme replacement therapy—was already helping some children, but it was costly and cumbersome. In the early 1980s, experiments used the enzyme taken from cows, but it worked only fleetingly before children became allergic to it. Then, in 1986, researchers discovered that linking cow ADA to polyethylene glycol—antifreeze—would keep it in the bloodstream long enough to prevent the killing of T cells. But "PEG-ADA" is hardly a cure-all. It only partially restores immunity. "T cell levels peak and then go down. And it costs about $250,000 per patient per year," says Donald Kohn, director of the Human Gene Medicine program at UCLA, who has worked on gene therapy for ADA deficiency for more than two decades, including with Anderson in the early days at NIH. Kohn, an energetic speaker with dark hair and a big bushy mustache, often talks about the history of gene therapy at scientific and medical meetings.

By 1984, the team that would carry out the first gene therapy for ADA deficiency was coming together. Kathy Anderson introduced her husband to Michael Blaese, chief of the cellular immunology section at the National Cancer Institute, who was an expert on the disease and had an office near French's, who now directed the molecular hematology branch of the National Heart, Lung, and Blood Institute. Blaese and Kohn had access to children with ADA, and they collected blood and bone marrow samples from two young patients in Wisconsin. Meanwhile, French Anderson had obtained the gene from one of the three researchers who isolated it. Kohn corrected T cells from one of the Wisconsin patients with normal ADA genes, then stitched the genes into mouse retroviruses stripped of the genes needed to replicate. What would happen when researchers reinfused such cells into the bloodstreams of young children?

Following his own advice to do animal experiments before moving on to people, Anderson and his colleagues at Memorial Sloan-Kettering Cancer Center in New York City tried the gene therapy on twenty-five monkeys with ADA deficiency. None showed corrected cells in their blood. The researchers were very discouraged, but Anderson implored them to try just one more monkey. And monkey #26—Robert—showed fixed cells in his blood. The saved simian kept the plan to try the gene therapy in children on track.

Years later, researchers discovered why the gene therapy had helped Robert the monkey. Unlike his lab-raised neighbors, Robert came from the wild, where he'd contracted malaria. He wasn't sick, but the infection had sent his bone marrow into overdrive, pouring cells into his bloodstream much faster than happened in the other monkeys. His revved-up circulation had sped the gene therapy to detectable levels. Robert the monkey unwittingly provided a clue that would prove important for gene therapy some twenty years later.

In 1986, Anderson submitted to the Recombinant DNA Advisory Committee a 500-page analysis of the cell and monkey experiments, ending with his appeal to begin treating children. But the committee's review wasn't what he'd hoped. Part of the reason for the RAC's reticence was the sheer strangeness of gene therapy, and the fact that the patient would be a child. Part was the politics of science, because Anderson had made some enemies. The largest factor, though, was that one lucky monkey wasn't strong enough evidence that the gene transfer would be safe.

Communications between the researchers and the RAC went back and forth. Then a way to get the first gene therapy trial off the ground came from an unexpected source—a new team member. The National Cancer Institute's chief of surgery, Steven Rosenberg, was making headlines with a new approach to treating cancer called immunotherapy. He removed white blood cells from patients with advanced melanoma and added genes whose protein products, interleukins, strengthen the immune response. The manipulated cells were then put back into the patients. It was a different disease, but the same approach. So in 1988 Blaese and Anderson sat down with Rosenberg to discuss collaboration, and the three instantly clicked. When Rosenberg pointed out to the RAC that several cancer patients die every minute, the committee members became more comfortable with the general idea of manipulating genes. Cancer was something they could relate to. Rather than a detour, the foray into cancer was a brilliant way to make the RAC pay attention to a rare disease.

The strategy worked. On May 22, 1989, a fifty-two-year-old melanoma patient from Indiana, Maurice Kuntz, with three months to live, had the gene treatment. Two days after an infusion of his own doctored white blood cells, corrected cells showed up in his tumors, and their numbers steadily rose. Kuntz lived nearly a year, and nine other patients lived longer, too, with no adverse effects.

Then the excitement began. Writing in *Science* that year, Theodore Friedmann called gene therapy "an unprecedented new approach to disease treatment through an attack directly on mutant genes" whose feasibility had already earned "broad medical and scientific acceptance." That wasn't completely accurate. Kuntz's immune system had been boosted, which wasn't the same as correcting a genetic flaw by replacing a gene. But it was a step in that direction.

At the quarterly RAC meeting on June 1, 1990, when ADA deficiency came up again for discussion, the committee asked the researchers about who should be treated. Using the sickest patients would give the clearest results, but was it ethical to deprive them of enzyme replacement therapy to do so? A compromise was to treat very sick kids, but also give them PEG-ADA. Blaese and Ken Culver, who'd replaced Don Kohn, were still seeking patients from their network of pediatric immunologists across the country, and a friend of Blaese's, Melvin Berger, referred two of his patients—Ashi and Cynthia. Both girls lived in Ohio.

The RAC decided to advise the FDA to approve the proposal. Because the buzz was that another rejection was looming, some of the key journalists following the story didn't bother to attend the final RAC meeting, nor did anyone else from the public. Even though by summer's end the FDA still hadn't given a thumbs-up, the researchers started taking Ashi's white blood cells anyway and adding working ADA genes to them. Ever optimistic, they wanted to have something to infuse into their first patient once the approval came.

The NIH was getting ready too. On September 3, anticipating imminent FDA approval for the trial, the agency held a press conference for Anderson to announce what would happen the next day. This time the media paid attention. Jeremy Rifkin, president of the Foundation on Economic Trends, known widely for objecting to biotechnology, showed up waving notice of a lawsuit filed because the final RAC decision had come when no one from the public was present. The press conference also set Anderson on the road to fame.

That night, final FDA approval for the trial still hadn't been given. Anderson barely slept, heading to the NIH clinical center at 5:30 the next morning to check on the cancer patients. At 8:55, the two agencies finally stopped bickering and the FDA gave its okay. Anderson waited with Ashi in the pediatric ICU. He took more blood, and then, with Ashi's hometown doctor, they watched Dumbo cartoons. At 12:52 p.m., Ken Culver brought in the "soup"—a pint of murky fluid containing about 10 billion of Ashi's corrected white blood cells—and attached the bag to her intravenous line. The infusion took twenty-eight minutes.

The NIH press office went wild, even though the results of the gene therapy weren't yet known. Blaese and Rosenberg avoided the limelight, as Anderson emerged as the media star. The NIH dubbed him the "father of gene therapy," which would become the subtitle of his biography.

Ashi had no side effects, and started improving after a few infusions. She received eleven treatments in all, one to two months apart, plus PEG-ADA every week. The regimen appeared to work. At the six-month mark, Ashi, her parents, and her two sisters all came down with the flu, and she was the first to recover. That, her mother said, was when they began to look at their daughter as normal.

Other clinical signs pointed to a mounting immunity. Her blood had more corrected T cells and more ADA. She also had positive skin tests for diphtheria, tetanus, and the fungus *Candida albicans*, which proved she was making antibodies to protect against infection. Plus, her treated cells were outliving the ADA-deficient ones, which was one major criterion for success. But how good was the science? Ashi's case presented a sample size of one, with no controls. Her taking PEG-ADA at the same time was a well-intentioned but scientifically confounding factor. Genes are highly regulated, transcribed into mRNA only when biochemical signals indicate that the encoded proteins are needed. Giving Ashi ADA as a supplement could stifle any corrected genes. "The enzyme replacement therapy may have blunted the selective advantage of the corrected cells. So the gene therapy was of minimal clinical benefit," says Kohn. In addition, he points out, because T cells normally live a short time, correcting them might provide only a brief effect, but one that was masked by the continuing ADA supplementation.

While Ashi was getting better, Anderson continued on his star trajectory, with awards, cover stories, honorary degrees, meeting the president, and being named runner-up for *Time*'s Man of the Year title. He was a consultant to the film *Gattaca*, and was even listed in a "heroes of medicine" article with the likes of Hippocrates, Edward Jenner, and Louis Pasteur.

Four months after Ashi's first gene transfer, the team treated nine-year-old Cynthia Cutshall. She had had a milder course of ADA deficiency, but was still a veteran of some serious infections, including septic arthritis at age five and painful sinus infections. She was diagnosed at six, after her T cell count plunged. She responded well to enzyme replacement for two years, but then her immunity failed. Cynthia never responded as well to the gene therapy as Ashi. For unknown reasons, only a tenth the number of viruses entered the older girl's cells.

The signs of real, lasting immunity continued. T cell counts rose. Both girls reacted to foreign antigens rubbed into their skin, and Cynthia grew tonsils and palpable lymph nodes, which she had never fully developed when she lacked immunity. Vaccines took hold. Ashi started kindergarten. Cynthia caught up in weight and height, and her headaches and sinus infections disappeared.

Ashi's and Cynthia's physicians were thrilled with the progress. Although a few people voiced concern about using retroviruses, which could theoretically activate cancer-causing oncogenes, to deliver genes, French Anderson and others continued expressing their enthusiasm. In *Science* in 1992, Anderson, who had just left NHLBI for the University of Southern California so that Kathy could take a surgery position, wrote, "human gene therapy has progressed from speculation to reality in a short time," and raised the possibility of abusing the technology for genetic enhancement. At first, he speculated, gene therapy would help thousands, and not millions, because of the special skills required, but he envisioned one day injecting gene-loaded vectors as easily as a person with diabetes injects insulin. A 1993 pamphlet from NHLBI, entitled "Curing Disease Through Human Gene Therapy," trumpeted: "With these men of diverse experience and talents working together, gene therapy became a reality much faster than it otherwise would have."

Although Ashi's and Cynthia's gene therapy took place over the course of two years, the NIH clinicians watched them closely for a long time. By 1995, both girls still had some corrected T cells and enough ADA to fight off infections. (The viruses must have entered rare stem or progenitor cells to have kept the correction going.) Ashi did remarkably well, and was often photographed with one of her physicians. She even spoke at scientific conferences, where she thanked the researchers for saving her life. I met her at a meeting when she was seventeen, and she told me about her plans to attend college and go into the music business. She has since earned a master's degree in public policy and married. Today, Ashi is healthy.

Despite the seeming success of the gene therapy for ADA deficiency in the two girls from Ohio, they weren't immune to everything. Perhaps this was because T cells are choosy about which germs they attack. Researchers began to think that the gene therapy might be longer-lasting if it fixed more pliant cells, such as the CD34+ cells found in bone marrow and umbilical cord blood. (CD34+ refers to a protein on the cell surfaces.) These cells enter the bloodstream and give rise to many others. The CD34+ bone marrow cells include stem cells and slightly more specialized progenitor cells. In March 1992, a group of investigators from Milan collected CD34+ bone marrow progenitor cells from a sick five-year-old boy, sent in healthy ADA genes, and gave them back to the boy. He did well.

In the spring of 1993, Donald Kohn and Kenneth Weinberg, at Children's Hospital, Los Angeles, had an even better idea—collect stem cells from the umbilical cord blood of newborns with ADA deficiency, and add the needed gene to those, so the correction would be present from infancy. They found three families who'd already had a child with ADA deficiency, were expecting the birth of a second, and agreed to participate.

On Tuesday, May 11, Andrew Gobea was born at Children's Hospital. His parents had lost their first child at five months. Andrew, with his shock of black hair, slept peacefully for the two minutes it took to infuse the gene-corrected umbilical cord stem cells. Three days after Andrew was born, Zachary Riggins was born at the University of California, San Francisco. His pediatrician collected the precious cord blood cells, sent them south for Don Kohn and Ken Weinberg to "fix," then got them back and infused them into Zachary within a week of his birth. Zach had a healthy four-year-old sister, and also a two-year-old brother being treated with PEG-ADA. Andrew and Zachary appeared on the cover of *Time* on June 24, under the headline "Brave New Babies." Both boys also received PEG-ADA—to be safe and to avert criticism from bioethicists. But receiving PEG-ADA obscured interpretation of the experimental results. If the children got better, how could anyone tell whether it was the replacement enzyme or the replacement gene?

The children treated for ADA deficiency with gene therapy continued to make a wonderful medical success story. The press ate it up. In 1995 Ron Crystal, chief of the division of pulmonary and critical care medicine at New York-Presbyterian/Cornell Medical Center, wrote in *Science*, "Once considered a fantasy that would not become reality for generations, human gene transfer moved from feasibility and safety studies in animals to clinical applications more rapidly than expected by even its most ardent supporters."

To this day, the gene therapy community isn't sure whether the gene transfer alone helped Ashi and Cynthia, Andrew and Zachary. We did learn something, Don Kohn says, because the girls suffered no ill effects, and still had marked T cells years later. Theodore Friedmann is more outspoken. "ADA deficiency was the first major disaster in gene therapy, not because it didn't work, but because it set a deceptive tone. Those were the early vectors, and we worked with what we had. But it was sold to scientists and the press as therapy." He claims there was never adequate evidence of gene transfer, let alone clinical benefit. "Those kids were never all that sick, not deathly ill. We were sold a story. The investigators acquiesced to inaccurate descriptions by the media. The Ashi DeSilva story was a figment of their imaginations."

Anderson disagrees. "It's uncertain if gene therapy helped Cynthia all that much, but it certainly saved Ashi. She still has approximately twenty percent gene-corrected T cells, in spite of the inhibitory effect of PEG-ADA."

Because of the uncertainty, some researchers have gone back to the drawing board, so to speak, for the ADA deficiency form of SCID. A paper in the August 22, 2011, issue of *Human Gene Therapy*, for example, describes using a different type of virus (adeno-associated virus, or AAV) to treat mice with the disease. Using mice, researchers can dissect and examine various tissues to learn where in the body the enzyme is made, which might provide insights into attempting the therapy in children. So even as gene therapy clinical trials inch ever forward, the preclinical research stages on animals are repeated and refined to provide more information.

The uncertainty of the first gene therapy, in which concomitant enzyme replacement therapy blurred the results, is in sharp contrast to Corey's story. His cure was due *only* to gene therapy.

8

SETBACKS

JESSE GELSINGER WAS NOT TO BE THE ONLY YOUNG PERson to lose his life in the pursuit of gene therapy. It was another form of SCID—the disease class that had been the first to be treated, probably successfully, with gene therapy—that contributed the next tragic chapter. The problem this time unfolded at a much slower rate than Jesse's four-day struggle. And again, from the ashes of an experiment gone horribly awry sprung new knowledge that led to the safety of Corey's treatment.

The next gene therapy to raise headlines by harming health was one that targeted SCID-X1, the "bubble boy" disease. In 1994, a decade after David Vetter died, Alain Fischer, a pediatric immunologist at Necker Hospital for Sick Children in Paris, developed a mouse with a knocked-out gene for SCID-X1. The animal was a model for the human version of the disease. By 1997 Fischer and his colleagues had learned enough to apply to the French government to begin a clinical trial to correct the condition in boys. Like ADA deficiency, SCID-X1 was a good candidate for gene therapy for a few reasons. A bone marrow transplant can cure it, which meant that correcting the right cells, in a way that gave them an advantage to replace sick cells, should restore a child's immunity. A dog model and a case report of a patient who got better when his gene mutated back to normalcy also suggested that gene therapy would work.

SCID-X1 cripples immunity differently than does ADA deficiency, Ashi's disease. Instead of a missing enzyme, boys who have

SCID-X1 have a defective or absent part of receptors that normally poke out of T cell surfaces, like little catcher's mitts. Healthy receptors on these T cells bind cytokines called interleukins, which trigger a variety of immune responses. SCID-X1 impairs part of these interleukin receptors, so the message to respond to infection can't get into the T cells.

The French researchers used retroviruses to carry the genes. These vectors can enter only actively dividing cells, and the targeted T cell progenitors undergo a brief frenzy of division before they mature. After that they live a fairly long time, quietly controlling all sorts of immune functions. So the researchers planned to slip the gene-loaded retroviruses into the cells that give rise to the mature T cells in the blood.

The research proceeded quickly. In 1999, Fischer's team gave an eleven-month-old and then an eight-month-old with SCID-X1 their own bone marrow cells that had picked up the engineered retroviruses outside their bodies, an *ex vivo* approach like the one used to treat Ashi. Using viruses in this way takes advantage of what viruses naturally do—they bind to protein receptors embedded in our cell membranes, then enter by either joining and traversing the membrane, or being "gulped" in by a process called endocytosis, in which the cell membrane envelops the virus. It is a little like an oil droplet either directly joining a larger blob of oil at the edge, or the larger blob snaking around and embracing the droplet.

Both boys stayed in the hospital for about three months. Neither had side effects, and their symptoms improved. The severe diarrhea and skin lesions that had plagued the younger boy vanished. Soon, T cells in the boys' blood bore the telltale marks of correction. The gene therapy appeared to be working.

Researchers typically wait several months to report the results of gene therapy trials, to be sure that the genes actually hit their targets, release their cargo, and correct the inherited error. The initial report on the first five boys treated for SCID-X1 didn't appear in *The New England Journal of Medicine* until April 18, 2002, two years after the first two boys were treated. At that time, they had no adverse effects or infections, a normal repertoire of T cells, and even their shrunken thymus glands had expanded. This was a key response, because the thymus gland churns out new T cells—in fact, the *T* stands for "thymus." In short, the researchers wrote, the experimental treatment "allowed patients to have a normal life." The gene therapy seemed safer than bone marrow or cord blood stem cell transplants, which carry the risk of graft-versus-host disease, in which transplanted tissue attacks the recipient's organs.

But the gene therapy did much more than restart the boys' immunity. It caused leukemia.

At the time *The New England Journal of Medicine* (*NEJM*) article appeared, Fischer's group had treated nine infants and one teen. Most of the patients were improving, but then one child developed what at first seemed innocuous—an elevated white blood cell count. When the boy's liver and spleen felt swollen to the touch, the doctors did a more complete blood analysis and found that the hiked white blood cell count reflected an increase in the numbers of a particular subtype of T cell. Such a sudden imbalance suggests rapid division from an ancestral cell—cancer.

By summer's end, the boy had roughly 300,000 lymphocytes (T and B cells) per milliliter of blood; normal is 10,000 to 40,000. Something was very clearly wrong, but the boy only had symptoms of mild anemia. By analyzing the DNA in the expanding group of extra cells, the researchers discovered that the gene-carrying viruses had inserted into a site on chromosome 11 that activates a nearby gene that causes acute lymphoblastic leukemia. Like the Henry Wadsworth Longfellow poem that reads "I shot an arrow into the air, it fell to Earth, I knew not where," retroviruses nestle into human DNA according to their own rules, and so triggering cancer had always been a possibility. (The virus in Corey's eye instead remains outside the chromosomes, quietly directing production of the RPE65 protein without jostling the resident genome.)

As the weeks went by, the boy's liver and spleen enlarged to handle the white blood cell overload, and he had chemotherapy. Fischer alerted other researchers and regulatory agencies right away, but he needed time to tell the parents of all the children in the clinical trial, who came from several European countries. The European officials were willing to keep quiet about the news for a few days, to give Fischer time to tell the families so they wouldn't hear it on the news. But in the United States, the RAC demanded public discussion right away, even after the French government attempted to intervene. So Fischer called the families on October 3, 2002, on the eve of the media frenzy. A headline in the *Los Angeles Times* on October 4 read, "Child 'Cured' by Gene Therapy Develops Cancer," damning an entire technology with punctuation. Despite the widespread coverage, including French Anderson telling *The New York Times,* "We knew it would happen sooner or later," the parents of the sick boy were very understanding, reassuring Fischer that they understood their son was in an experiment.

The researchers in the gene therapy community, still reeling from Jesse Gelsinger's death, were growing very uneasy. Was the boy with leukemia a fluke? On December 2, a second case was discovered. Both boys were under three months of age when treated.

The FDA learned about the second boy on December 20, and put a hold on clinical trials using retroviruses to transfer genes. Embarrassingly for the French researchers, their brief report of the first case in the January 16, 2003, *NEJM* suggested that the boy's recent chickenpox or family history of cancer might explain his leukemia. The next day, when *Science* published a news article about the second case, the French government placed the gene therapy trial on hold, and the United States stopped twenty-seven more gene therapy trials.

In October 2004, one of the first two boys to develop leukemia died of it, and by January 2005, a third child, a nine-month-old treated in the summer of 2002, had it. All in all, the French group had treated ten children, and a group in the UK had treated another ten. Of those twenty boys, the SCID improved in eighteen. But five boys developed leukemia.

Researchers are still debating what went wrong in the French SCID-X1 clinical trial. "A lot of factors caused it, and we still don't know for sure what happened," says Don Kohn. But it is clear that several risk factors converged. The patients were tiny and very young, with poor immunity, and got huge doses of a virus whose genetic material had a natural tendency to zero in on an oncogene.

Despite the setback, efforts continued to develop gene therapies for ADA deficiency and SCID-X1. Don Kohn, who'd treated the prenatally diagnosed Andrew Gobea and Zachary Riggins in 1993, started another trial in 2001 using two different retroviruses to deliver ADA. The proposal, submitted two years earlier, had wallowed in the regulatory morass in the wake of the Gelsinger tragedy. But soon after the trial began, researchers in Milan successfully treated two infants who had ADA deficiency with a different protocol: The patients were *not* given the supplementary ADA enzyme—the PEG-ADA that had obscured the earlier trial results—but *did* receive a chemotherapy drug, busulfan, which makes room in the bone marrow for corrected cells to accumulate. Remembering Robert, the monkey who'd responded to gene therapy for ADA deficiency when malaria increased the capacity of his bone marrow, Kohn revamped his trial to add the chemo and take away the ADA. But he had to wait until 2005 to enroll six more patients because, in the United States, the FDA had placed ADA deficiency gene therapy trials on hold, in light of the leukemia cases.

When clinical trials resumed, researchers checked blood counts carefully. They also administered lower doses and treated only children older than six months, because the affected boys in the French trial had been so young. The children also had to be younger than five years, which is when the thymus, which houses many T cells, begins to shrink. Informed consent documents were rewritten to point out the possibility of leukemia should the vectors go astray.

This new-and-improved ADA deficiency gene therapy worked. "We now have patients making their own T cells with the gene and not needing the enzyme," Kohn says. The Milan and UK groups have altogether treated a few dozen patients with ADA deficiency, most of whom have responded. And because it has worked, chances are that the 1990 gene therapy on Ashi and Cynthia did, too.

Gene therapy for SCID-X1 continues to reinvent itself. In a new trial treating boys in London, Paris, Boston, Los Angeles, and Cincinnati, researchers are using a different virus: HIV, stripped of the genes that cause AIDS. In the United States, the RAC had strict recommendations: the boys must be at least three years old, unable to find a stem cell or bone marrow donor, or be so sick that they wouldn't survive a transplant.

The children successfully treated for the inherited immune deficiencies who are today making headlines could perhaps be regarded as the children who saved gene therapy, but the timing, I think, points more toward Corey. The SCID cases began *before* Jesse Gelsinger's death derailed the field, but it took years to assess their success, and the medical journal papers describing that success appeared *after* I read about Corey for the first time in my hometown newspaper in November 2008. His story touched me in a way that the others, which I'd chronicled in magazine articles and in my human genetics textbook, didn't, for the circumstances were very different. Recognizing the efficacy of gene therapy for both types of SCID required years of observing the absence of something—infections. In contrast, Corey's response was both sudden and stunning, restoring a biological function easily imagined—sight. In addition, media coverage of SCID-X1 focused, for a time, on the leukemia cases. The gene therapy for LCA2 had no such taint to overcome, and made a great story for reporters even before they heard about Corey, for the treatment had first worked on several young adults and, before them, on sheepdogs.

The extraordinary impact of the successful gene therapy for LCA2 transcends my fascination with one boy. In August 2011, the National Institutes of Health announced a new program in its "Common Fund," which Congress enacted in 2006 to encourage within-agency collaborations. The new program is in "Gene-Based Therapeutics: Manipulating the Output of the Genome to Treat Disease." The stated goal: "One or more gene-based therapies would become established as a treatment option for patients with genetic disease. As a benchmark, gene-based therapy would become as common as bone marrow transplantation is currently at major academic medical centers. Such an outcome could transform the clinical outlook and lives of patients with genetic disease. This would be especially important for rare diseases where in most cases no other treatment options exist." The first reference cited in the program summary is Corey's clinical trial.

A look at the tragedies in gene therapy that led up to Corey Haas's success would not be complete without considering Jolee Mohr. Her gene therapy is an important part of the overall story because it may have worked *too* well.

Jolee didn't have an inherited condition, like Corey, Jesse, or Ashi, but she did have a disease of the immune system, like the SCID kids. She had rheumatoid arthritis (RA), a classic autoimmune disease in which the immune system attacks the body, affecting about 1 percent of the world adult population, including 1.3 million people in the United States.

Jolee spent all her life in Illinois. The joint swellings of rheumatoid arthritis began in 1993, when she was twenty-two years old, causing great knee pain. For eight years she tried standard drug treatments, including prednisone (a steroid that fights inflammation) and methotrexate (which limits cell division). In 2002, treatments became available that target a protein with the unsightly name tumor necrosis factor (TNF), which causes inflammation. These newer drugs—Enbrel, Remicade, and Humira—block TNF from binding to its receptors on the cells that line joints. They are protein-based drugs that patients must inject, because if ingested, digestive juices would rip them apart.

Jolee started on Enbrel, then switched to Humira in 2004, but her knee was so painful that she kept taking the old drugs, too. None of them, in any combination, really worked—the pain in Jolee's right knee never totally went away. So in February 2007, when her rheumatologist at the Arthritis Center in Springfield, Illinois, suggested that she consider enrolling in a clinical trial for a gene therapy that might bring more lasting relief, she and her husband, Robb, were excited. It was a whole new way of attacking the disease. Injecting a gene that would quell inflammation right at its source might help, the doctor suggested, and unlike a conventional drug, it might do the trick without a daily dose.

It was an intriguing idea. Dr. Robert Trapp had cared for Jolee for seven years, so she trusted him. She and Robb went home to read the fifteen-page informed consent document and think it over. Dr. Trapp had already signed up a few of his patients and received a small fee from the drug company for recruiting them, which is common practice. It was a phase 1/2 clinical trial, testing an escalating dose in small patient groups. The goal was to assess safety, but using doses that could be effective.

The gene therapy would deliver the same DNA sequence that is the basis of the drug Enbrel, encoding a bit of protein that would plug up the TNF receptors, which would block the inflammation signal. It would go right into the inflamed joints of trial participants, aboard adeno-associated virus as the vector. Researchers had learned a great deal about the tiny AAV since adenovirus (AV) was sent into Jesse's liver, and it appeared to be much safer. AAV doesn't cause disease in humans and in fact is already present in many of us, so a strong immune response, like the one that killed Jesse, is extremely unlikely. AAV is also unlikely to insert itself into an oncogene and cause cancer, as happened to the SCID-X1 kids. Manufactured by the Seattle-based Targeted Genetics, by 2007 AAV had already been used to safely shuttle genes in nearly fifty clinical trials.

The rationale behind the gene therapy for rheumatoid arthritis made sense. Despite the name tumor necrosis factor, which conjures up images of a rotting cancer, TNF triggers inflammation, which may be a good or a bad thing, depending on circumstance. If the body is infected, inflammation makes conditions inhospitable for germs, washing them away. But if the body doesn't need the inflammation, as in Jolee's case, the result is painful autoimmunity.

Jolee, her husband later said, didn't pay much attention to the jargon-riddled consent form, and thought the experimental procedure would help her pain. She may have missed the mention of possible adverse effects, including "pain, discomfort, disability or, in rare circumstances, death." A note in her medical chart indicates that Dr. Trapp answered all of her questions about the trial.

The procedure seemed straightforward: two shots in her right knee. The first would be either the gene therapy or a placebo (a sham shot of empty viruses), and the second shot, twelve to thirty weeks later, would be the treatment. The trial had been under way for two years already. Of the 127 participants, 74 had received their second dose, and responded so well that, in March 2006, the company had increased the dose to 50 trillion viruses per milliliter of joint fluid, in the knee, ankle, wrist, knuckle, or elbow. Targeted Genetics planned to report the promising results at the next American College of Rheumatology annual meeting in November.

On February 26, 2007, Jolee's doctor administered the gene therapy to her right knee. She felt so little that she thought she'd received the placebo, she told Robb. But the aftermath of the second shot, which she got on Monday, July 2, was another story. The parallels to what happened to Jesse Gelsinger, in retrospect, are chilling.

Jolee took her last dose of Humira at the end of June, and that weekend, she, Robb, and their five-year-old daughter, Toree, went boating. Jolee was a little under the weather, more tired than usual after a week at her job entering data at the secretary of state's office. By Monday she was still tired, but went to Dr. Trapp's office for the second shot anyway. She had a temperature of 99.6 degrees.

By Tuesday, Jolee felt much worse. She was throwing up and running to the bathroom with diarrhea, and her temperature had risen to 101 degrees. By Wednesday she was still sick, so Robb took her to their family doctor, who diagnosed a viral infection. The next day, he put her on antibiotics, in case it was bacterial. But by Saturday, July 7, her condition had grown alarming. Her temperature was now 104 degrees and still rising, so Robb took her to the local emergency department. She was diagnosed with an infection and liver damage and sent home. Meanwhile, the family physician called Dr. Trapp, who mentioned the gene therapy trial, insisting that Targeted Genetics had said it was safe. Both physicians were puzzled by their patient's condition. By the next Monday, Jolee's heart rate was dangerously accelerated and her white blood count soared, despite a flood of antibiotics. Her liver function tests were very abnormal, and her right knee was hurting again.

When Robb took his desperately ill wife back to the emergency department the following Thursday, July 12, she still had symptoms of an infection. Her bloodwork didn't reveal any bacteria or viruses, but beneath the radar of the tests a fungal infection was stealthily beginning to brew. At the time, all that the doctors could tell was that Jolee's liver was failing, and she was now having difficulty breathing. Scans revealed a thickened gallbladder, enlarged liver and spleen, and fluid in her abdomen. Then her blood pressure plummeted as she began to bleed internally so profusely that transfusions couldn't keep up. With her hemoglobin level and blood pressure down, her liver enzymes up, and her abdomen swelling, it was becoming clear that Jolee's ravaged body had sprung a leak. But even as late as July 17, no one connected this alarming clinical picture to the trillions of altered viruses injected into her body the week before.

In the ICU, Jolee's health continued to spiral downward, and her physicians worried that her liver might completely fail. A liver transplant was the only option. So on the night of July 18, the local hospital physicians sent her by ambulance two hundred miles to the University of Chicago Medical Center. Jolee arrived shortly after midnight and went straight to the ICU, as the search for a liver went into high gear. Finally, an ICU physician learned that Jolee had recently received gene therapy, and called the FDA. The next day, July 20, the FDA put the gene therapy trial on official hold.

Meanwhile, Jolee's condition was worsening. Not only was she on dialysis for kidney failure and a ventilator for respiratory failure, but the blood pooling in her abdomen now ballooned out and pressed in, crushing her organs. Still, no one knew what was causing Jolee's massive internal bleeding. The team in the Chicago ICU learned that she wasn't very sick when she received the two gene therapy shots—both at the highest allowed doses. But they were puzzled because samples of her liver tissue examined under an electron microscope didn't reveal escaped gene therapy viruses. A blood smear taken on July 21, however, held a hint—yeast cells that had burrowed into Jolee's white blood cells. Yeasts are single-celled fungi.

Finally, even mechanical ventilation failed and on July 23 Robb Mohr tearfully signed a "do not resuscitate" order for his wife, since her brain was no longer receiving oxygen and the damage was irreversible. Treatments stopped, and she was heavily sedated, for comfort. The bag of blood that was her abdomen was by now so hideous that Robb wouldn't let little Toree see her mother. The next day, as the end neared, he couldn't bear to be in the room. The doctors turned off the life-sustaining equipment, and Jolee Mohr died, twenty-two days after receiving a shot of gene therapy to her knee.

But it wasn't the gene therapy that killed her.

Preliminary autopsy results, available by July 26, were shocking. Jolee's blood was swarming with fungi, as were her liver, lungs, bone marrow, spleen, lymph nodes, pancreas, brain, and intestines. The organisms had probably nicked a blood vessel in her abdomen, causing the slow leak that fed the bulging eight-pound bag of blood. Analysis of her stored blood, using highly sensitive technology, revealed that the fungus that causes the infection histoplasmosis had in fact been lurking in her body, albeit in low numbers, on July 2, the fateful day when she'd received the second gene therapy shot. Since Jolee had spent her entire life in an area where the fungus was endemic, it's likely that her extreme immune suppression—from using Humira as well as the gene therapy—had reactivated an old infection.

The direct cause of death was *Histoplasma capsulatum*, a fungus found in soil, caves, and bird and bat droppings in the Ohio River Valley and the lower Mississippi areas. In more than 99 percent of people who inhale fungal spores, the immune system prevents or squelches lung infection. But if a person with a healthy immune system harbors a low-level infection without knowing it, and then the immune system is suppressed, the infection can suddenly rage out of control. People taking immunosuppressant drugs and living in an area where the fungus is endemic face such a risk.

As the parallels to the Gelsinger case mounted, the FDA and Targeted Genetics launched a volley of news releases. The media, sniffing another gene therapy death, arrived like vultures to collect the gory details and relay the story. *The Washington Post* led the pack again, with an August 6, 2007, article outraged anew at the ethics of a clinical trial not intended to help the individual, although this is the very definition of *phase 1.* The media were guilty, again, of therapeutic misconception. And this is where the two tales of gene therapy deaths diverge. Gene therapy caused Jesse's death, but in Jolee's case, the evidence pointed in another direction, a reawakened infection.

Targeted Genetics defended the gene therapy right from the start. Two days after Jolee died, the president and chief executive officer at the time, H. Stewart Parker, announced, "The clinical course that this individual experienced has, to our knowledge, never been seen as a consequence of exposure to adeno-associated viral (AAV) vectors or naturally occurring AAV." On September 17, 2007, the RAC agreed, concluding that it was "unlikely that the AAV injection was a significant contributing factor to the patient's death." Attention began to shift to the other immunosuppressant medications she'd been taking before and during the gene therapy.

On November 11, Philip J. Mease, MD, clinical professor at the University of Washington School of Medicine and the lead investigator in Jolee's clinical trial, reported that the gene therapy itself appeared to be not only safe but effective. He also evaluated the molecular tests on Jolee's tissues that looked for spread of the virus, concluding that the vector "did not contribute to the patient's death, which was due to disseminated histoplasmosis and a severe retroperitoneal hematoma"—referring to the fungus and the bag of blood. Sixteen days later, the FDA lifted the hold on the gene therapy trial, and another RAC meeting on December 3 placed the blame on Humira. With this news and continuing patient reports of reduced arthritis pain, Targeted Genetics began planning the phase 2 trial—but this time, to be safe, they disqualified patients who had been on immunosuppressants. The full autopsy results on Jolee Mohr weren't reported in a medical journal until two years after her death, but they, too, exonerated the gene therapy.

The combined dampening of Jolee's immune response, from the three preexisting treatments and then the gene therapy, opened up her body to an overwhelming fungal infection. Warnings about blocking TNF were on the packages of the drugs she had taken when she received the gene therapy. On the day that she died, the package insert for Humira had a "black box" warning highlighting the elevated risk for opportunistic infections: "Many of the serious infections have occurred in patients on concomitant immunosuppressant therapy that, in addition to their rheumatoid arthritis, could predispose them to infections."

Bioethicists weighed in, en masse, on Jolee's death in the January 2008 issue of *Human Gene Therapy*. Art Caplan explained again the problem of therapeutic misconception, the idea that a clinical trial is meant to treat or cure. It persists. When I asked Alan Milstein, the attorney who represented Paul Gelsinger and then Robb Mohr, in 2010 how he would advise a client about participating in a gene therapy trial, he said, "What was wrong with Gelsinger was that he was a healthy volunteer. Jolee had a mild case of RA. If somebody was severely ill and had no other options, so that the risk and benefits were even, then depending on the trial and the research, it may be worth submitting to that kind of experiment. Unless there really is no other option, I can't see participating."

The sad stories of Jesse Gelsinger, the boys with SCID-X1 who developed leukemia, and Jolee Mohr make Corey Haas's success all the more remarkable. The failures taught powerful lessons to those who plan and carry out gene transfer experiments, in pursuit of gene therapies. The ADA deficiency experiments showed researchers to target less mature white blood cells that could better amplify and sustain the response, while leukemia in the SCID-X1 boys revealed the need for more research on the habits of the viruses commandeered as gene therapy vectors. In addition to selecting participants more carefully and paying attention to every nuance of health in the weeks preceding vector administration and at every point thereafter, researchers became alert to the most difficult types of problems to spot—errors of omission, such as not knowing that Jesse Gelsinger had had neonatal jaundice, and that the immunosuppressant drugs that Jolee Mohr took could allow a lurking infection to explode.

Jesse, Jolee, and the others laid the groundwork for the successful gene therapy on the left eye of Corey Haas. Another giant step forward in gene therapy came in late 2009, with a report on the successful treatment of a terrible disease that slowly sends bright, vibrant boys into vegetative states: adrenoleukodystrophy (ALD). The world first heard of ALD from the story of a beautiful dark-haired boy named Lorenzo, and the oil mixture that his parents invented to try to save him. Effective gene therapy for ALD is now possible thanks to the family of another beautiful boy who had the disease. His name was Oliver.

9

LORENZO AND OLIVER

PARENTS WILL DO ANYTHING FOR THEIR CHILDREN. After learning that their child has an untreatable inherited disease, some parents decide to take matters into their own hands; they are desperate to do *something*. They frantically read the medical and scientific literature, googling and zipping through PubMed using all the relevant key words and phrases, trying to follow the intersecting cycles and pathways of biochemistry to understand their child's disease. They brainstorm what seems to be a logical treatment, or find someone who can. Petrified parents, diagnosis finally in hand, may turn to alternative therapies, from massage or magnets in a mattress, to all manner of extracts and elixirs, diets, vitamins, and nutritional supplements.

Many a desperately researching parent has discovered the story of an astute mother in Norway who noticed that her children's diapers had a strange odor. Her observations, told to intrigued experts, led to one of the greatest success stories in medicine. The treatment the experts developed compensated for a missing enzyme with a special diet. With much effort on the part of the affected child, the approach worked, spectacularly. But it illustrated in a profound way, like the new drugs for cystic fibrosis, that manipulating a biochemical other than DNA, the underlying genetic instructions, isn't a forever fix.

The story of PKU began in 1931, when the Norwegian mother

of two young disabled children noted a musty odor to their urine. The father mentioned this to a friend, who told a friend, who happened to be a physician with an interest in biochemistry. Intrigued, the physician analyzed the foul urine in a lab at the University of Oslo, with help from the mother, who hauled over bucketsful in the course of a few weeks. The physician, Asbjörn Fölling, not only identified the inherited problem in the children's metabolism—later named phenylketonuria, or PKU—but went on to find it among hundreds of people languishing in mental institutions. Because of a missing or nonworking enzyme, their bodies couldn't convert the amino acid phenylalanine into another, tyrosine. The phenylalanine built up, poisoning parts of the brain. Since phenylalanine is a normal part of protein, eating fewer protein-rich foods seemed sensible—and this dietary strategy worked, if begun early enough, and prevented mental retardation and other symptoms. For a few years, photos of PKU families showed the eldest children, born prediet, sitting in wheelchairs and staring blankly out into space, next to their animated, healthy younger siblings, who'd also inherited PKU but had followed the diet.

At first physicians instructed parents to keep their kids on the low-protein regimen until six years of age, when the brain was thought to stop developing. The diet was (and is) rather unpalatable—bland, white, specially formulated "medical foods," making it hard to get kids to comply. Then parents and physicians noticed that after older kids began to eat normally, their mental acuity dimmed. When the first group of people with PKU who followed the diet for six years began having children of their own, disaster struck. All the children of PKU mothers were mentally retarded—not just those who'd inherited PKU. It turned out that after the diet stops, even if the patient seems healthy, phenylalanine is still building up. In pregnant women, the buildup poisons the fetus. So today, people with PKU restrict their diets for life. For decades, all newborns in the United States have been tested for PKU with a blood sample from the heel. If the telltale biochemical signs are there, the diet starts immediately.

Treating an inherited disease with a special diet was a first step toward gene therapy, because it corrects the condition after a mutant gene acts. The 1992 feature film *Lorenzo's Oil* beautifully illustrated this approach toward ALD, adrenoleukodystrophy.

Lorenzo Odone was born on May 29, 1978, to Augusto, an economist with the World Bank, and Michaela, a linguist. They lived in Fairfax, Virginia, an easy commute to their jobs in the nation's capital. Augusto still lives there today. In the summer of 1983, Augusto was sent to the Comoros Islands, an archipelago of volcanic origin off the southeastern coast of Africa, west of Madagascar and east of Mozambique. Michaela and Lorenzo came along. It was an idyllic time, remembers Augusto. "Lorenzo learned French in the Comoros and some Comorian words. He was a very gifted, precocious child." He also knew English and a bit of his father's Italian. Lorenzo was stunning, even better-looking than the actor who would play him in the film. He loved classical music and Greek mythology.

The boy was healthy, active, and strong while in the Comoros, but became ill when the family returned to the United States. Kindergarten started out well, but then Lorenzo began to have difficulty paying attention. The usually calm child would fly into rages and break rules. At Christmastime he fell, and by the spring, he was falling so often that it was clear he couldn't see. Lorenzo started to have blackouts and memory loss. He had difficulty speaking, had seizures, and was very tired.

Michaela and Augusto took their son to Children's Hospital in Washington, where physicians ruled out a brain tumor, seizure disorder, Lyme disease, attention-deficit/hyperactivity disorder (ADHD), and just clumsiness. Then white spots on his brain MRI indicated adrenoleukodystrophy. The protective layers of fatty myelin around the neurons were disappearing. Symptoms of ALD typically begin between the ages of four and eight and rapidly worsen. Lorenzo would die by age eight, predicted the experts with somber certainty. Bone marrow transplants were just starting to be used to treat ALD, but the challenges were daunting. Lorenzo would need to find a matched donor. The procedure was extremely risky, and it wouldn't reverse the damage already done.

Lorenzo's brain scan explained the angry outbursts in school—the bright boy was frustrated with his rapidly vanishing skills. He could hear, but his speech now sounded like gibberish. He lost his ability to write the letters of the alphabet and could no longer sound out words. The doctors said he would never learn to read. His spatial orientation, already off, would worsen, so that he would lose the ability to walk. Visual disturbances would fade to blindness, while seizures might become more frequent. Lorenzo's aggression and unruly behavior might worsen and then ebb as he withdrew into himself, away from the world. Eventually, he'd lose control of his muscles and become demented. It was about as devastating a diagnosis as a parent can imagine.

ALD affects only 1 in every 18,000 to 21,000 boys, so it wasn't surprising that the Odones had never heard of the disease. But now they at least knew the enemy. Michaela and Augusto got to work right away, huddling over biochemistry, molecular biology, and genetics journals and textbooks at the NIH library in Bethesda. They learned that their son's condition had first been described in 1923. Family studies in 1963 revealed an X-linked recessive pattern of inheritance, the disease passing from carrier mother to affected son with a risk of 50 percent. (Very rarely, women have symptoms.) But not everyone who inherits the mutation develops symptoms. No one knows why this is so, but it probably reflects the actions of other genes. Making things even more complicated, some people have a mild form (adrenomyeloneuropathy) that doesn't cause symptoms until after age ten. After that, these can range from lower leg weakness that lasts a normal lifetime to rapidly fatal brain degeneration.

In 1976, researchers studied the brains of boys who'd died of ALD to try to figure out what goes wrong. They discovered that in certain cells of the brain, adrenal glands, skin, and blood, fatty molecules called very long chain fatty acids—VLCFAs—build up to levels much greater than normal. The problem lies in tiny sacs in cells called peroxisomes, which are stuffed with enzymes and other important molecules.

Normally, the tiny peroxisome sacs have gateways, like portholes in a ship's side, made of pairs of a protein called ABCD1. The gateways let in an enzyme that breaks down the VLCFAs. The ALD mutation, in a gene called *ABCD1*, destroys the gateways. As a result, the fatty cargo builds up outside the peroxisome sacs. The fatty VLCFAs are normally cut into pieces of a certain size that the cells then combine with other fats and some proteins to make myelin, the insulation on brain neurons that enables them to transmit messages fast enough to keep us alive. Myelin isn't directly slathered onto the twists and turns of neuron branches. Instead, it fills supportive cells called neuroglia that encase neurons like Bubble Wrap. The lack of myelin, resulting from the lack of VLCFA pieces, deflates the Bubble Wrap, causing the symptoms of ALD as the neurons beneath, healthy but hampered, can no longer send messages quickly enough. Because the recipe for myelin is so complex, several such "demyelinating" conditions exist. The most common one is multiple sclerosis.

The abnormal brain cells found in people with ALD are a type of neuroglia called microglia. The cells that give rise to these neuroglia come from the bone marrow via the bloodstream. This fortuitous fact makes gene therapy for ALD simpler than for other demyelinating conditions, because the therapy is deliverable into the bloodstream. Gene therapies for other brain disorders must be introduced directly into the brain.

Hugo Moser, the "father" of ALD, who led a team at the Kennedy Krieger Institute in Baltimore until his death in 2007, discovered that the levels of VLCFAs in the blood rise before symptoms begin. He headed a large study on the brothers of boys who had ALD. Those who went on to develop ALD all had elevated VLCFAs before symptoms began. This meant that excess VLCFAs could be used as a marker of impending illness. Some of the symptoms of ALD—weakness, fatigue, stomach upset, weight loss, and darkening of the skin—come from the VLCFA buildup in the adrenal glands, which sit atop the kidneys and secrete several types of hormones. Fortunately, corticosteroid drugs help with these symptoms.

The Odones' library research eventually took them to the PKU diet. Could there be a dietary intervention for ALD? Adding nutrition tomes and journals to their reading, the distraught parents now paid closer attention to the defect in fatty acid metabolism that was melting the insulation off their son's brain cells. Nutrition was something they could influence.

Fatty acids are the building blocks of fats. They are long chains of carbon atoms fringed with hydrogens, their chemical formulae strings of Cs and Hs. Their unwieldy Latin names and chemical abbreviations reflect the number of carbon atoms. The fatty acids that build up in ALD are C26 and C24. In ALD, the VLCFAs are made too fast and broken down much too slowly. Most important, though, the bulk of them come from the diet. So as was the case with PKU, restricting their intake could theoretically balance things out, perhaps slowing, tempering, or even preventing the terrible symptoms.

In 1982, in the library stacks of the NIH, the Odones discovered that researchers had already tried to treat ALD by restricting the C26 and C24 fatty acids in the diet. It didn't work. Then in 1986, researchers found that adding oleic acid—a C18—sops up the enzyme needed to make the two guilty fatty acids, lowering their levels. Further digging led the Odones to a biochemist in England who created, on their advice, what would become known as Lorenzo's oil. It combined C18 and C22 fatty acids from canola, olive, and mustard seed oils.

Augusto joined Hugo Moser and his wife, Ann, and others to scientifically test the oil. In 1987 they published in the *Annals of Neurology* results of their "new dietary therapy" on thirty-six patients. Two years later they reported a refined recipe that reduces levels of VLCFAs within a month. But would lowering the VLCFAs prevent or relieve the symptoms of ALD?

As Lorenzo grew, the doctors' predictions came true. By age seven, the inquisitive dark-haired boy who loved classical music was bedridden and blind, able to communicate only by blinking and moving his fingers. Augusto and Michaela moved his bed into the living room, where he could be the center of attention, and encouraged their friends to visit and talk to him. Augusto retired from his position at the World Bank early, to devote himself to his son.

The Odones gave Lorenzo the oil, but his brain was already too damaged to respond. In that year, 1989, Augusto and Michaela started The Myelin Project, "a multinational gathering of families struck by one demyelinating disease or another." They funded the first international conference on ALD. The goals of The Myelin Project are very clearly stated on the website, spelling out that the Odones were in search of treatment or cure, not kudos to scientists: "Basic research and studies directed toward the advancement of science for science's sake are excluded from Project financing." The disconnect between the goals of basic research and clinical medicine in the eyes of a suffering parent was the same observation that a shattered Jim Wilson had come to a few years earlier, when he realized that his discovery of the mutation behind Lesch-Nyhan syndrome would not help his young patient Edwin.

While the Mosers were following the boys taking Lorenzo's oil, *Lorenzo's Oil* opened and brought instant attention to ALD. The 1992 film starred Susan Sarandon, who got an Oscar nod as Michaela Odone, and Nick Nolte as Augusto, with Zack O'Malley Greenburg in the title role. Sarandon took Michaela to the Academy Awards.

The multitalented Augusto Odone, economist-turned-scientist, had a hand in the film, as co-screenwriter. He tells how it happened. "George Miller, an Australian doctor who was also a director and producer, was having breakfast in Sydney, and he happened to see an article about the oil in the Sunday *London Times* and thought, 'If this story is true, I'm going to make a movie.' So he called me from Australia and asked if I would be interested and I said, 'Why not?' He was enchanted with the idea of the heroes, and he made a very nice film, about eighty percent of which is true and twenty percent is Hollywood," Odone recalls. Sure enough, tiny print at the end of the film, after all the credits and copyright notices have run, admits that the film is "a true story although certain characters and incidents are fictional."

Hugo Moser was not quite as enthusiastic about the movie as Augusto Odone. Dr. Moser wrote in *The Lancet* in 1993, "As a work of fiction, *Lorenzo's Oil* is an excellent film. However, as a factual documentary it has three main flaws." It exaggerates clinical trial results, invents conflicts between the Odones and their doctors, and does not accurately depict the intentions and activities of the United Leukodystrophy Foundation, Moser wrote. What most researchers who knew anything about the disease objected to, however, was the Hollywood ending of kids taking the oil looking much too healthy, although the Lorenzo character is still depicted as sick. Physicians and bioethicists blasted the film for raising false hopes, while biology instructors developed wonderful lesson plans based on the movie, and students as well as the rest of the public loved it.

Lorenzo's Oil faded from the headlines, but not from the medical journals. Studies were showing that lowered VLCFA levels in the blood were not a sign of the oil working, but just a marker. The level could go down, but brain damage continued on its relentless course—the symptoms didn't go away.

In 1993, Patrick Aubourg and Nathalie Cartier and their colleagues at INSERM (National Institute of Health and Medical Research) and the Hospital Saint-Vincent in Paris published in *The New England Journal of Medicine* their findings on using Lorenzo's oil to treat the milder, later-onset form of ALD. These researchers—who would go on to develop the successful gene therapy for ALD—studied the mild form because they thought that the severe childhood form progressed too fast for the oil to work. They evaluated symptoms, did MRI brain scans, tested nerve conduction, and measured blood VLCFA levels for up to four years in fourteen affected men, five symptomatic women, and five boys who had a mutation but no symptoms. Although the VLCFA levels sank down to near normal within ten weeks on the oil, no one improved. Nine of the fourteen men actually got worse, and one of the boys developed symptoms. It was a painful lesson for researchers and families alike: correction at a molecular level is not enough. A treatment has to alleviate symptoms as well as slowing or stopping disease progression.

Despite these disappointing results, parents continued giving their sons Lorenzo's oil, hoping for a small improvement or at least a delay in symptom progression. Then in July 2005 *The Journal of the American Medical Association* published the long-awaited follow-up of the Lorenzo's oil study, with Hugo Moser as first author and Augusto Odone the last, a position of honor. This study was crucial because it tracked 89 boys who had no symptoms when they started the oil, but had elevated VLCFAs and brothers who had severe cases of ALD. Each boy had taken Lorenzo's oil and followed a low-fat diet for seven years. Twenty-one of them developed brain abnormalities seen on MRI scans, and ten of them also developed neurological symptoms. There was no control group, but the researchers compared the results with an older study of 443 boys who knew they had inherited a mutation causing ALD, but were untreated because the oil hadn't been invented yet. About a third of them developed symptoms before age ten. Although comparing the studies was a little bit of apples-and-oranges, a greater percentage of the untreated boys went on to develop symptoms than did the boys who took the oil. The researchers tentatively concluded, "We recommend Lorenzo's oil therapy in asymptomatic boys with X-linked ALD who have normal brain MRI results." The Myelin Project website was more enthusiastic: "They found an oil that stopped the disease in its tracks." But taking the oil is not risk-free. About a third of treated boys develop a clotting problem, and deficiencies of certain fatty acids can arise.

The jury is still out on whether the oil works, because it is impossible to tell, for any individual, whether it was the oil that prevented symptoms, or if he would have remained symptom-free even without taking the oil. Still, Lorenzo's unexpected longevity suggests that the oil might have helped.

Lorenzo matured into a handsome man who resembled Sir Paul McCartney, with lush brown hair and exquisite brown eyes, his beard and light mustache a testament to his normally functioning endocrine system. Augusto and Michaela tried hard to maintain some normalcy for him. A squad of caregivers and nurses dressed and undressed him daily, frequently moved his useless limbs, and put him into a chair so that he could listen to music. An aide bathed him daily in a pool built off the living room. The serene look captured on Lorenzo's face in photographs when he was bathing perhaps meant that he was calm—or perhaps meant nothing at all. He wore diapers and was fed through a gastrointestinal tube five times a day, and had to be suctioned often, sometimes every few minutes, because he couldn't swallow his saliva. When Michaela died in 2002 of cancer, a family friend took over her tireless care of Lorenzo.

On May 30, 2008, Lorenzo choked and then bled to death—possibly because of the blood-thinning effects of the oil—before a rescue squad could arrive. It was the day after his thirtieth birthday.

Lorenzo Odone lived twenty-two years longer than anyone had predicted. Was the oil responsible? Today, even Augusto admits to having doubts. "It could have been his care *and* the oil. His mother was very, very careful, but even so, the oil had something to do with it." He pauses for a long moment. "But I'm not sure about that."

~

The Odones brought ALD into the public eye. The Salzman sisters—a social worker, a pharmaceutical industry executive, and a veterinarian—assembled the research team that developed successful gene therapy for the disease. I met Eve Salzman Lapin at the 2010 annual meeting of the American Society of Gene and Cell Therapy in Washington, D.C., on a warm May evening. Nervous about speaking in the morning to the media about her family's experience with ALD, she was happy to practice her talk on me.

Eve and Bobby, who live in Houston, Texas, decided to start a family shortly after they married. Oliver was born in 1992, the year that *Lorenzo's Oil* hit the big screen. Within three years, he had brothers, Elliott and Alec. The handsome, dark-haired boys were great friends. "The other two boys would follow Oliver everywhere.

If Oliver fell, so did Elliott, just to copy him," recalled Eve, a willowy brunette with a sad smile. She's glad that she left her social-work job shortly after Oliver's birth to stay home with him.

For a few years, the household was like any other run by three rambunctious boys. Oliver was very bright. "By age three he would do a puzzle of the United States that had the capitals written on the backs of the pieces. He memorized the state capitals and shapes. If you held up a puzzle piece, Oliver knew what state it was by the shape," Eve recalled.

Oliver loved preschool and kindergarten. But in first grade, just like Lorenzo, he began to have trouble sitting still and paying attention. A developmental pediatrician diagnosed ADHD and prescribed Ritalin. "We talked to parents of kids who really had ADHD and they said Ritalin helped. But it didn't help Oliver at all. The doctor increased the dose, and it still didn't help," Eve said. The next diagnosis-du-jour was Asperger's syndrome, a mild form of autism that seemed to fit Oliver's social awkwardness. "He went to a social skills group for Asperger's after school a few times a week. That taught him some things, but Asperger's wasn't the problem," Eve said. His memory was failing, and his once-neat handwriting grew disorganized and messy.

Finally, a psychologist helped the Lapins put the pieces together and referred them to a neurologist who ordered a brain MRI. The scan revealed the characteristic white matter destruction of ALD, already so far advanced that the doctor was astounded that Oliver's skill level was as high as it was. Eve and Bobby, who had thought their little boy had a learning disability, instead found out that he had a fatal brain disease.

If the diagnosis of ALD wasn't bad enough, it is also inherited, passed from carrier mothers to sons with a probability of 50 percent. Eve worried about her sisters, Amber and Rachel. Were they carriers too? Even as Eve and Bobby were still reeling from the bad news, everyone in the family who could have inherited the errant gene was tested. The results were grim. The middle son, Elliott, had the mutation, but Alec didn't. Rachel was spared, but Amber was a

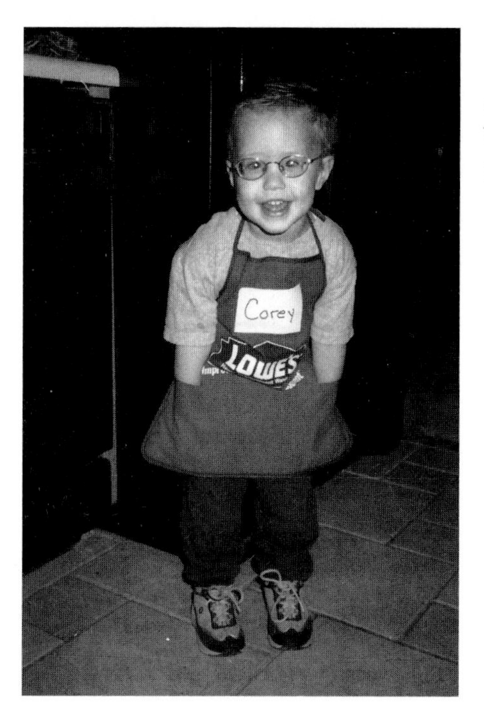

Corey was a clumsy toddler, but cheerful and curious. *(Ethan Haas)*

In late September 2008, four days after gene therapy for Leber congenital amaurosis type 2 (LCA2), Corey visited the Liberty Bell, still using his cane. The next day, at the Philadelphia Zoo, the bright sunshine hurt—for the first time ever. *(Ethan Haas)*

Jesse Gelsinger died on September 17, 1999, four days after receiving gene therapy for ornithine transcarbamylase (OTC) deficiency. He was eighteen. *(Migdalia Gelsinger)*

James Wilson, MD, PhD, is a professor of pathology and laboratory medicine at the University of Pennsylvania. He has developed many viral gene vectors.

Eve Lapin, Rachel Salzman, and Amber Salzman were pivotal in planning successful gene therapy for adrenoleukodystrophy (ALD), although their own boys would not benefit from it. *(Steve Barsh)*

Oliver Lapin *(center)* died a day before his twelfth birthday in 2004, of adrenoleukodystrophy (ALD). His brother Elliott *(right)* also inherited the disease, but a stem cell transplant helped him. Alec *(left)* is healthy. *(Alan Ross)*

Hannah Sames, age seven. She is one of fifty-one children and young adults in the world known to have giant axonal neuropathy (GAN). *(Wendy Josephs)*

The neuroscientist Paola Leone, PhD, has devoted her life to helping families with Canavan disease. *(Chris Herder Photography)*

Max Randell *(right)* had gene therapy for Canavan disease when he was eleven months old and when he was four. He loves his little brother, Alex.

Lindsay Karlin and her dad, Roger, celebrate their shared birthday, July 18, 2010. Lindsay turned sixteen. She was the first person to receive gene therapy for Canavan disease, in March 1996, shortly before her second birthday.

Kristina Narfstrom, DVM, PhD, began studying blindness in Briard sheepdogs in 1988. She poses with Pluto, who is mostly Briard (a great-grandmother was a beagle). Pluto had successful gene therapy for Corey's disease when he was four years old. (Today's Breeder *magazine, Nestle Purina PetCare)*

Lancelot was the first dog cured of LCA2 with gene therapy. He demonstrated his newfound vision to Congress at a luncheon—and didn't pee on the oriental rug. *(Foundation Fighting Blindness)*

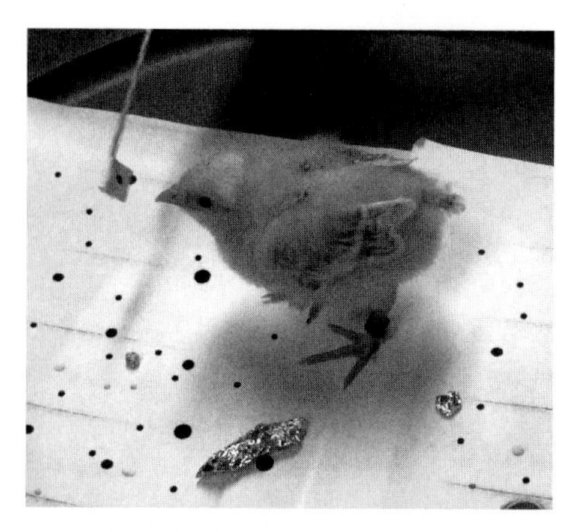

To show that gene therapy restores vision, humans read eye charts, dogs spin, and chicks peck at dots or candies. *(Susan Semple-Rowland)*

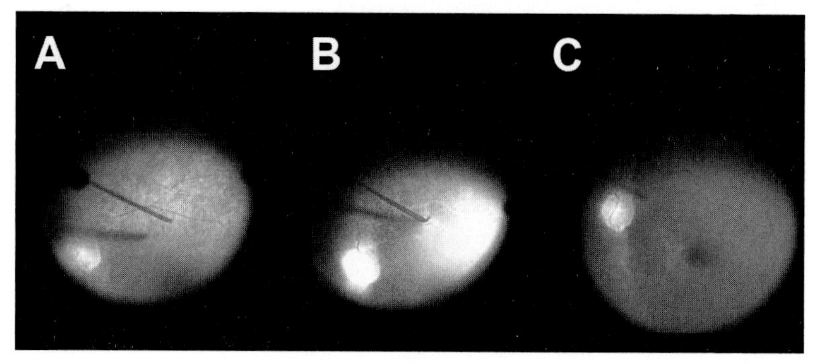

Gene therapy restores vision. In panel A, the cannula bearing the healing viruses nears the retina. In panel B, it makes contact and the injection of viruses beneath the retina begins. In panel C, Dr. Albert Maguire withdraws the cannula. *(Dr. Jean Bennett)*

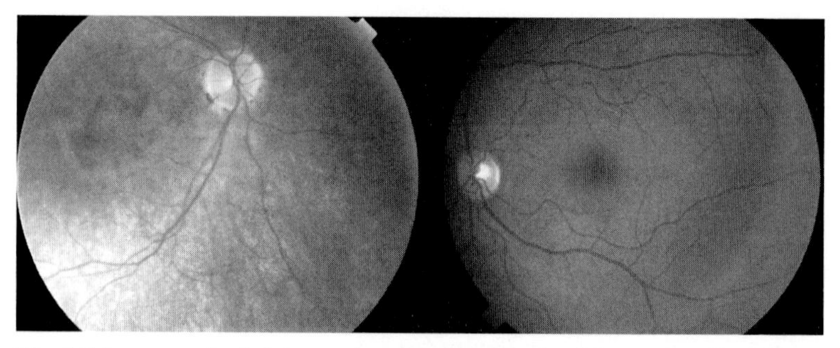

The LCA eye on the left is much paler than the healthy eye on the right, especially the dark central disc (the macula), which is the site of the densest photo receptors. The LCA eye also has a paltry blood supply. *(Dr. Jean Bennett)*

Katherine High, MD, director of the Center for Cellular and Molecular Therapeutics at the Children's Hospital of Philadelphia, holds a tube of viruses engineered to deliver human genes to cure blindness. Albert Maguire, MD, professor of ophthalmology at the University of Pennsylvania, operated on Corey, and Jean Bennett, MD, also a professor of ophthalmology, led the gene therapy trial. *(Children's Hospital of Philadelphia)*

"I'd like to introduce the youngest person ever to speak at ASGCT—this is Corey Haas," Dr. Jean Bennett told an astonished crowd of gene researchers at the 2010 annual meeting of the American Society of Gene and Cell Therapy in Washington, D.C. Corey sits next to his parents, Nancy and Ethan. *(American Society of Gene and Cell Therapy)*

Corey would never have been able to do this, without falling, if his LCA2 had progressed. *(Wendy Josephs)*

The Haas family: Ethan, Corey, and Nancy. *(Wendy Josephs)*

carrier like Eve, and her one-year-old son, Spencer, had the mutation too. It seemed incomprehensible that two young brothers and their baby cousin could all have the same dreadful disease. Amazingly, the sisters jumped into action, proceeding in two directions—treating the boys and at the same time starting the Houston-based Stop ALD, a not-for-profit organization to advance research.

The first step was to see if Oliver was still healthy enough for a bone marrow stem cell transplant. The tests of his skill levels were heartbreaking for Eve to watch. "He had trouble doing a four-piece puzzle. At age three he would have done it exactly right." But she also noticed something that was strangely comforting. "It didn't seem to bother him! I realized then that he was losing touch with reality, and I was glad he didn't seem concerned. It was a blessing for him, and that was a saving grace for me also." Oliver was too sick to benefit from Lorenzo's oil or a bone marrow transplant. However, a transplant might help the younger boys, whose brain scans did not yet show white matter destruction. So the Lapins held the largest bone marrow donation drive in Houston history. Although several people found a match, neither Elliot nor Spencer did.

Like Lorenzo, Oliver stayed at home, remaining very much a part of the family. But it was tough. "He lost one ability after another. Cognition went first, then speaking, seeing, walking, and the ability to move on his own. Eventually he lost the ability to swallow," said Eve in a quiet voice, lost in the painful past and going beyond what she would tell the room full of reporters the next day.

Eve and Bobby worried about the effect of Oliver's illness on his brothers. Elliott seemed remarkably well adjusted. When a psychologist asked if he felt guilty—which made Eve cringe—the boy easily replied, "Why would I feel guilty? I didn't make Oliver sick." The youngest brother, Alec, was eerily perceptive. "He was five when Oliver was diagnosed, so he doesn't really remember the Oliver that we knew. He only remembers Oliver losing abilities and being sick. He hears the stories we tell, and sees the puzzle of the United States, which we kept," said Eve.

Oliver was very sick those last two years. Every day, Eve and

Bobby bathed him and gave him muscle relaxants to prevent painful spasms, and suppositories to prevent constipation. He had physical therapy often, to prevent muscle atrophy and bedsores, because he could no longer move. Eve could tell from his facial expressions that he wasn't in pain and that he wanted her there. Sometimes she'd pretend that he was a newborn, eager to hear her voice but not yet able to respond in kind. "I tried to communicate with blinks, but it was more like an unspoken communication, a love communication," Eve remembered.

Finally, Oliver went on hospice home care. Each day, a member of the hospice team would visit—usually a nurse and sometimes a social worker. Special others came too, such as a family friend who gave Oliver massages. The hospice nurse gently explained to Eve and Bobby exactly what would happen as the end neared for their son, and the network of ALD experts that Amber and Rachel were assembling also helped. Eve related the events with a stoic calm.

"When he passed away we were all there with him. I decided I wanted to hold Oliver all the way to the funeral home. Not in a hearse, but in a van. Bobby and Elliott stayed home. But Alec, I don't know if it was compassion or curiosity, wanted to help.

"'Why don't I go with you?' asked Alec, who was eight.

"All I could do was cry. But Alec continued.

"'Do you think . . . do you think Oliver is comfortable? Is he in pain?'

"'No, he's not in any pain anymore. He's comfortable.'

"'Then why are you crying?'

"'I'm crying for me, not Oliver.'

"He was relieved to know Oliver was not in pain."

Oliver Lapin died a day before his twelfth birthday.

The special tragedy of a genetic disease is that it can strike more than once in a family. Just when Oliver began to decline faster, Elliott's brain MRI showed demyelination, the beginning of the stripping of the neuronal insulation, and a scan a few months later revealed spreading white matter abnormality. The clock was ticking. A transplant was now imperative to stall the disease in the second son.

While Eve and Bobby were caring for Oliver and reexploring a transplant for Elliott, his cousin Spencer had a brain MRI that also showed the first inklings of the disease. But a new option had recently become available for children, like the cousins, who couldn't find a bone marrow match—a transplant of stem cells from umbilical cord blood. Umbilical cord stem cells are not as festooned with surface features that could irk a recipient's immune system as bone marrow stem cells from an adult donor, and so might help the boys even if the stem cells did not perfectly match their cell surfaces. It was worth a try.

Cord blood stem cell transplants were still experimental in 2002. The procedure required a long hospital stay, so the two families rented houses next door to each other near Duke University Hospital in Durham, North Carolina. At Duke, Joanne Kurtzberg, chief of pediatric blood and bone marrow transplantation, was pioneering the approach. Alec stayed in Texas with Bobby's parents. Amber, her husband, Steve, and baby Spencer relocated temporarily from Philadelphia.

The transplant would require up to three months in isolation, because the boys' immune systems would be wiped out. Then several months of frequent checkups and tests would track the transplanted stem cells as they traveled in the boys' bloodstreams and traversed the blood-brain barrier, slipping between the tilelike cells that form the tiniest blood vessels of the brain. Everyone hoped the procedure would seed the boys' brains with healthy cells that would put myelin back on the neurons, stalling the disease.

Going to the hospital for the transplant was very strange, because Elliott and Spencer weren't really sick—yet. "Many of the kids on the ward had leukemia, and they were sick and going into the transplant because chemo hadn't worked. So here were these kids, bald and weak, and they saw Elliott and said, 'Why are you here?'" Eve remembered. Elliott understood why he needed the treatment, from seeing what Oliver had gone through. But the family was unprepared for the side effects of the stem cell transplant. Most people think of a transplant as not working when the recipient's immune system rejects the "foreign" tissue. But the opposite happens too. The transplanted bone marrow or cord blood stem cells can attack the recipient's body because the cells are, in essence, a foreign immune system that "sees" itself as the host, not the invader. The result is graft-versus-host disease, or GVHD.

Spencer was lucky. His transplant stopped his ALD without sparking rejection or GVHD, and today you'd never know he had a deadly genetic disease. Elliott didn't fare as well. The transplant halted his ALD, but short-term side effects blossomed into the long-term disability of GVHD, just when his older brother was dying. Elliott suffered terribly right from the beginning of his treatment.

The three days of intense chemo used to destroy Elliott's bone marrow in preparation for the transplant sucked the energy out of him. He became listless from nonstop vomiting, developed a high fever, and broke out in hives. Even if he'd had an appetite, he wouldn't have been able to eat because his entire digestive tract was inflamed. "It was excruciating for him to eat or drink. For the duration of the time in the hospital, Elliott was attached from tubes in his chest to an eighty-pound pole pumping fluids and drugs through his frail and weak body. He took a thousand doses of medicine a month, each with short- and long-term risks. We were prisoners of the unit since his lack of immunity required specially filtered air," Eve recalled. But within a few days of the transplants, the cousins' white blood cell counts were inching up—the transplants were working.

Elliott lived in the transplant unit for three months, as graft-versus-host disease ravaged his small body. The acute, usually short-lived form of GVHD typically starts within the first three months of the transplant, as it had with Elliott, but once he returned to the rented temporary home and the onslaught continued, it became clear that he had chronic GVHD, bringing the fear that the immune attack could spread to his central nervous system. His aunt Amber cried as she recalled her nephew's distress. "They came home from the clinic, and Elliott sat on the toilet, throwing up and with diarrhea, with a bucket in his lap and passing out. If he had had gene therapy, what he suffered could have been avoided." Months after these symptoms finally abated, Elliott began having difficulty walking. The GVHD had spread to his brain and spinal cord after all. Elliott began to use a wheelchair just after Oliver died, and he's had several surgeries since. But his ALD has stopped.

Even without rejection or GVHD, the aftermath of a stem cell transplant can be brutal, as Spencer found out. "Dr. Kurtzberg said Spencer's transplant was easy. But that still meant living in the hospital for a few months, tied to that pole, and having several touch-and-go moments. Compared to other kids, though, Spencer did okay. He had no long-term complications," said Amber.

When genetic testing revealed that Spencer had inherited the ALD mutation, Amber and Steve entered the strange world of "assisted reproductive technologies." They had wanted more children all along, but now they had a directed goal: to have a child who was free of the mutation and also a tissue match for Spencer. Since the boy had been diagnosed at a year old, doctors thought he would need another transplant in four or five years. For some conditions, stem cell transplants must be repeated to sustain the effect.

To conceive a baby who would be a tissue match for Spencer, Amber and Steve turned to IVF-PGD—*in vitro* fertilization with preimplantation genetic diagnosis. At the time PGD was still a relatively new technique. An egg is fertilized in a lab dish and a few cell divisions are allowed to ensue, then a cell is plucked from the tiny, early embryo—a ball of cells—and tested for the family's mutation. The rest of the embryo remains viable, so if the mutation isn't there, that embryo is implanted into the woman, and nature takes over for the rest of the pregnancy.

It had been a little more than a year since the first "savior sibling" had been born. Adam Nash was selected and gestated to save his six-year-old sister, Molly, from Fanconi anemia. The Nashes made the talk-show rounds, despite receiving much criticism. (In 2004, the novelist Jodi Picoult borrowed from their story in *My Sister's Keeper*, which depicted a teen suing her parents for intentionally conceiving her to take parts of her to treat her sister's leukemia.) Today, PGD is often done along with IVF, and is even used frivolously, such as to balance the numbers of boys and girls in a family.

Today Amber laughs about the efforts to conceive her daughter Lane, who was finally born, after a stint in an embryo freezer, in 2003. "If you go to an IVF clinic and you are over forty, they look at you as if you are a thousand. 'Your odds are terrible!' they warned me, but I had five cycles of IVF. I figured if any embryos were ALD, we'd donate them for research. If they didn't have ALD but were not a match for Spencer, we'd freeze them."

The first attempt to get an ALD-free, Spencer-matched embryo didn't take. Time was running out, and they tried again. This time, Amber was in New York City having her embryos retrieved, and they would have to be flown to Wayne State University in Detroit, to the PGD guru Mark Hughes. But the date was September 11, 2001. "So my husband, a recreational pilot, drove to New Jersey with embryos in hand, and flew them to Michigan," Amber recalls.

Dr. Hughes typed, selected, and froze several embryos. The rest was up to Amber's uterus. Three times, embryos failed to implant. The fourth attempt became Lane, who was born on August 13, 2003—Oliver's eleventh birthday. "It was Oliver who helped make it possible for us to have a healthy baby. We attribute Lane's strong will and personality to surviving after a cell was taken from her as an embryo and then being in the freezer for a couple of years. She is a tall, strong, athletic girl who still never wears a jacket," says Amber, laughing. Lane is not a carrier—both of her copies of the ALD gene are normal. Amber and Steve did not select carrier embryos because they didn't want a daughter to have to deal with passing on the disease to her sons. Someday, if Spencer's stem cell transplant needs a boost, Lane can help.

While Eve and Bobby dealt with their two boys, Rachel and Amber were active with the Stop ALD Foundation. Amber looks very much like Eve, tall and thin with shoulder-length, slightly wavy brown hair, but Rachel has a much longer and darker mane, reminiscent of the tails of her equine patients.

Like the Odones fifteen years earlier, the sisters read the scientific literature and spoke to ALD experts around the world. "There weren't very many, so we approached others who might contribute ideas. We brought doctors, scientists, and researchers from many disciplines together who helped us get over many obstacles," said Eve. One giant obstacle was the then recent Jesse Gelsinger tragedy.

The Salzman sisters used their unusual collection of skills and connections to pursue the tainted technology. Rachel offered the deep science background of a veterinarian, while Amber had a PhD in mathematics and was a senior vice president at the pharmaceutical giant GlaxoSmithKline (GSK). (Today she is president and CEO of the biopharma company Cardiokine, based in Philadelphia.) She explained how they started Stop ALD. "When my nephew was diagnosed, we did a lot of research, and what we learned was dismal. It was December 2000, and I called my boss at the time, Tachi Yamada, the head of R and D at GSK, and said I needed help. And Tachi said, 'Let me call Jim. You need to pursue gene therapy.'"

Jim was Jim Wilson, who at the time was the focus of the investigation looking into Jesse Gelsinger's death.

Amber, who worked in the Philadelphia area, met Jim, and they had their first discussion with Rachel chiming in on the phone from Florida. "I didn't know anything, and Jim was so patient. He laid out exactly what we'd have to understand to get to a clinical trial. It was all overwhelming, and it didn't happen overnight," Amber recalled.

The gene therapy approach was straightforward, and the success of stem cell transplants indicated that if replacing cells worked, so would replacing genes. The strategy would actually combine the two. Correct the right bone marrow cells outside the body, put them back in, and they'd travel in the bloodstream to the brain. There they'd gradually replace the microglia that had the plugged peroxisomes and, slowly, restore the myelin sheaths on denuded brain neurons. To turn that fairy tale into reality, Stop ALD would need a transplant surgeon, a vector person, an ALD expert, and a neurologist. "We also needed people in the rare disease community, and the FDA. Jim got Don Kohn and others in the gene therapy field, and we hosted a meeting at GSK" in March 2001, Amber recalled. An early decision was to replace the inefficient retroviruses used in many animal studies and some clinical trials with a specific type of retrovirus, called a lentivirus. These viruses offer several advantages as vectors: they readily enter many cell types, carry large loads, and don't bother the immune system if tweaked. The earlier retroviruses were the ones that caused leukemia in the SCID-X1 boys.

Jim Wilson coached Rachel and Amber in how to cut through egos and alliances to assemble the best team. They recruited Hugo Moser, the father of ALD, and Patrick Aubourg and Nathalie Cartier, of INSERM, the public research institute in Paris. "Jim was the catalyst. I called him a lot, with technical and political questions. How do you play the game?" Amber recalls. And he fondly remembered their enthusiasm. "Rachel and Amber were always there. In Europe, in the U.S., they'd take us out to lunch, be in the hallway, the hotel lobby. They did everything shy of following us into the men's room."

The gene therapy trial for ALD came together as Rachel and Amber jetted all over the world, meeting and introducing the key players. They rendezvoused in London with Inder Verma of the Salk Institute in La Jolla, who was championing the use of a lentivirus called HIV. They tracked down Gabor Veres, who developed gene therapy vectors for Cell GeneSys, a California biotech company. He was at a meeting in Washington, en route to Hungary for another meeting, when he got a call from Amber, who introduced herself and then implored him to meet with Aubourg and Cartier. During the flight, Veres received a message from Amber from the flight attendant: she'd already arranged for Patrick Aubourg to meet him in Budapest. Rachel flew out to introduce the two scientists. A stellar team emerged.

Jim Wilson is still in awe of the sisters. "They were pushing an aggressive agenda at a time when everyone was in full retreat." When the meeting at GSK took place in March 2001, the memory of Jesse Gelsinger's death in 1999 was still fresh, and the vector used to treat two boys for SCID-X1 had already silently harpooned a cancer-causing gene. As Amber and Rachel continued to assemble their gene therapy team for ALD, the first boys treated for SCID-X1 developed leukemia, and the field detonated. "The first leukemia case was a disruption; with the second, the RAC shut down trials. But they never gave up. Amber and Rachel eliminated the unseemly parts of science—the competition, the gamesmanship. They were smart and earned credibility with the scientists and physicians, and that is how they completely disarmed us."

The French researchers were eager to start the trial, so Amber and Rachel had to do a lot of convincing to complete the team, in light of the ongoing gene therapy controversy. Amber eventually got the CEO of Cell GeneSys, Steve Sherwin, on board. "Sherwin wanted to focus on oncology. The last thing he needed was an adverse event in a gene therapy trial. So we held meetings, had conversations, and finally the company agreed to provide the vector for the trial in France," recalled Amber. The vector cost Cell GeneSys more than $1 million.

Convincing families to enroll their sons with ALD in a gene therapy trial that uses HIV, when other gene therapy trials had recently killed people, was challenging, and the first patient didn't sign on until 2006. Then business decisions intervened: Cell GeneSys was absorbed into another company, the ALD project was dropped, and Veres, the all-important vectorologist, was let go. But the Salzman sisters convinced the new company to make the vector anyway.

The idea to tame and turn HIV into a gene therapy tool dates back to 1998; Inder Verma of the Salk Institute patented the use of the virus as a vector in 2000. Unlike other retroviruses, a lentivirus such as HIV can infect a nondividing cell, and typically takes years to produce symptoms, compared with other viruses that make their presence known in weeks. Jim Wilson recalled that the first time Patrick Aubourg showed other scientists evidence of HIV's efficiency in delivering genes, getting more cargo inside cells than other vectors, jaws dropped. "It was what everyone had dreamed of. It worked on so many cell lineages, it would open the door to any disease. Of course there was the fear that the public would think we were giving people HIV, but the virus is disabled front, left, and center. It is impossible to generate HIV." Even so, researchers tend to refer to the popular vector as "lentivirus," and not "HIV," especially when speaking to the media.

Timing is important in ALD gene therapy. To work, gene therapy would have to intervene early, before the lack of myelin irreversibly damages neurons. Using gene therapy too late in the game might lock someone into the state that Lorenzo and Oliver were in toward the end, for potentially a normal life span because the disease wouldn't progress. Developing gene therapy to treat ALD was also necessary because stem cell transplants have limitations. For ALD, they take too long to stop the disease: twelve to eighteen months. "The cells go from the bone marrow to the blood to the brain, but replacement of brain microglia is slow, and you need to replace a certain percentage for clinical benefit," explained Nathalie Cartier at the ASGCT meeting in Washington. Meanwhile, damage continues. Stem cell transplants from donors are also risky, causing 20 percent mortality in children and up to 40 percent in adults. "That's why we proposed gene therapy using the patient's own cells," she added. Gene therapy would replace cells faster than a stem cell transplant because DNA sequences that control the expression of the needed gene could be knitted right into the virus.

The clinical trial for ALD began at Saint Vincent de Paul Hospital in Paris, right around the time that Corey's doctors were presenting their case before the RAC to use gene therapy to treat children with LCA2. The first boy with ALD was treated in France in September 2006, and the second in January 2007. Patient 1 was four years old and patient 2 three and a half at the time of the gene transfer. Neither had bone marrow or cord blood matches. By the time that patients 1 and 2 made headlines, two others had been treated.

The boys didn't feel anything much different than did little Ashi DeSilva back in 1990 when she received the first gene therapy for an inherited disease. First, a drug "mobilized" certain stem cells in their bone marrow to slip into the bloodstream. The researchers then sampled blood and separated out the needed cell types—those destined to home to the brain—and then exposed those cells to lentiviruses carrying healthy *ABCD1* genes. The researchers had also linked a "reporter" gene to the therapeutic gene, marking cells that picked up the payload with pink dots, so cells with the transferred gene would stand out. After a few days, the cells with the pink dots were making the needed protein. As the precious fixed cells grew in lab dishes, the boys took chemotherapy drugs to clear out their diseased bone marrow to make room for the new, as Don Kohn had done with children who had ADA deficiency. Finally, each boy received about 5 million of his own doctored cells.

The results were excellent, although not as immediate as Corey regaining his sight days after treatment. The gene therapy for ALD worked better and faster than stem cell transplants. Still, wary from the SCID-X1 trials on the European boys, the researchers waited many months before publishing the results on ALD. They wanted to be certain that what they were seeing was real, and safe. It was.

There was no leukemia, no HIV infection, no rejection, and no graft-versus-host disease. And the evidence that the gene therapy was working rang out at all different levels.

Blood tests tracked a steady increase in the proportion of white blood cells bearing the fixed gene, including their precursors in the bone marrow, as levels of very-long-chain fatty acids in the blood fell. This meant that the cells were not only dividing, but producing what they were supposed to. Brain scans showed a halt in the melting away of the fatty myelin and heads slowly filling with new white matter. Signs were clear in the boys, too. Patient 1's cognitive skills improved, although he is still in special education classes because of a poor attention span. Patient 2 stabilized and attends regular school, and patient 3, treated at twenty-eight months of age, has no neurological symptoms at all and is cognitively normal. All three boys became able to make the enzyme that they had been born unable to produce. In addition, only about 15 percent of the microglia were corrected. In the many conventional stem cell transplants that the team had done, correction of 80 percent had been necessary.

The gene therapy for ALD indeed "stops the disease in its tracks," as was once claimed for Lorenzo's oil. The Salzman sisters as well as the researchers are already thinking ahead to the next goal: teaming the gene therapy with newborn screening, to treat boys with ALD right from the start. It is an idea that would work for Corey's disease too, and many others. Says Eve Lapin, "The legacy of Oliver's life and death is that gene therapy will be a better way to treat ALD and other terrible diseases."

The boys combating ALD with gene therapy were treated after Jesse Gelsinger's death and after leukemia developed in the SCID-X1 boys, and so they, too, helped to save gene therapy. In fact, the paper reporting the ALD success appeared in *Science* on November 6, 2009, just a day before the *Lancet* paper announcing the results of Corey's clinical trial, making it appear in the headlines as if the return of the biotechnology was a double overnight breakthrough.

What the ALD story adds to the evolution of gene therapy, and to the seeming miracle of what happened to Corey, is the parent perspective. Nancy and Ethan Haas felt the worry, shock, despair, and helplessness of facing an inherited illness in a child. But they were lucky. They never had to haul stinky diapers to a biochemist to figure out why their children were mentally retarded. They didn't have to spend hours in the library teaching themselves biochemistry and nutrition to invent an oil that might halt the destruction of their son's brain. They didn't conceive one child to save another. And Corey's parents didn't have to form a not-for-profit organization or fly around the world convincing reticent researchers to collaborate to cure a rare disease. As the gene therapy successes continue, more and more parents may find themselves in the Haases' position—easily entering a clinical trial.

The family sagas of developing gene therapies for two other diseases provide insights into the start and end of the journey that Corey is now on—the initial efforts to start clinical trials (giant axonal neuropathy, or GAN) and living with the aftermath of gene therapy (Canavan disease).

PART IV

BEFORE GENE THERAPY

Mommy, why do you have to go
away to learn science?

—Hannah Sames, age six

10

HANNAH

ON A DRIZZLY EARLY SPRING DAY IN 2010, LORI SAMES and her vivacious daughter Hannah drove from their home in Rexford, New York, up state route 50 toward Ballston Spa, avoiding the Adirondack Northway, which transports tourists from New York City to Saratoga Springs or Lake George. Few vacationers venture off the upstate highways to explore the picturesque towns in between. Lori and Hannah passed a dairy farm on the right, with the Green Mountains of Vermont painting the distance in the rainy haze. Hannah looked out the window, hoping to see the cows.

Ballston Spa has the feel of another era. The McDonald's just south of the antique-store-lined main street is the only modern brand name. Novelists including Caleb Carr, William Kennedy, and Richard Russo have evoked the timelessness of the town in their books. In *Nobody's Fool*, Russo calls the town "North Bath," but people who live in the capital district of New York easily recognize it as Ballston Spa. The name commemorates the natural mineral spring "bubbling up cool and delicious from the low swampy ground . . ." according to an 1878 history of the town. The Indians had ignored the spring, favoring the waters in nearby Saratoga, but surveyors discovered it in 1771 and spread the word. (Today, the faintly sulfurous odor of the local springwater makes one wonder what the fuss was about. The forest near the water smells like a toilet.)

On a two-block-long side road parallel to Milton Avenue, the main street in Ballston Spa, a small, cream-colored, maroon-trimmed Victorian house sits behind a sign that reads "Living Well," and in smaller print "Healing Arts Center & Spa." Lori asked me to meet her and Hannah here for a visit to a naturopath. Since conventional doctors hadn't been able to help Hannah much, and most of them had never even heard of the little girl's disease, Lori wanted to explore what complementary practitioners might have to offer.

I went up a short flight of wooden stairs to a porch, where a bulletin board was adorned with local business cards and announcements. In the tiny town, one can find an expert in Reiki, Pilates, nutrition, or even "interspecies telepathic animal communication," whatever that is. I was a little early, so I ventured inside for a peek. No one was around, but in this part of the state, we don't usually lock our doors.

Bells tinkled when the front door opened onto a waiting area that had the feel of a pampering spa. The room was elegantly appointed, with comfortable chairs and a delicate floral scent just a few aerosolized molecules short of being cloying. Brochures on tiny tables advertised such services as Tui Na Oriental Bodywork, glycolytic peels, and a variety of facials offering concoctions of collagen, algae, rose, vegetable mud, seaweed, or herbs to be massaged into parched epidermis, or minerals for those with a taste for the inorganic. Steep stairs to the side of the entranceway led to offices and exam rooms of various types of health care practitioners.

I was back on the porch perusing the bulletin board a few minutes later when Lori pulled her minivan into one of the diagonal parking spots. She helped Hannah out and they headed toward the porch stairs. Lori has a fresh-faced look, with just a touch of frazzled mom. It takes a moment to notice that something is wrong with Hannah. She's adorable, with extremely kinky hair and a bright smile. That day she was clad in jeans and a jeans jacket with a pink shirt that echoed the pink barrette holding some of her frizz. Because her legs are slightly bowed, she had difficulty hauling herself up the short flight of stairs, but that and her hair were the only obvious unusual features. By the time school started in September, Lori whispered, Hannah would need a walker, and soon after, a wheelchair.

Hannah was here today to address her frequent stomachaches. She's tiny—a mere forty pounds—with a voracious and odd appetite. That morning, she'd had asparagus, dill pickles, and coleslaw for breakfast. Lori doesn't know if her daughter's penchant for strange foods is part of her inherited illness, or just Hannah.

Mother and daughter were there to meet with Sarah LoBisco, a naturopath. Hannah also regularly sees speech, physical, and occupational therapists, a psychologist, a chiropractor, a physiotherapist, a neurologist, an orthopedic surgeon, a gastroenterologist, and of course a pediatrician. None of them know much about her condition, but they try to help manage her symptoms.

Dr. LoBisco, an attractive, energetic brunette, came downstairs. Poised and friendly, she had an instant rapport with Hannah, who was sprawled on the floor, working a puzzle. We went upstairs, slowly, past two rooms with therapeutic-looking beds in the centers, to an office in the back. Hannah spread her puzzle on the floor while Lori and I drew up chairs. Dr. LoBisco sat at her desk and opened her laptop. A diploma above her indicated that she has a four-year graduate degree from one of only four institutions in the United States certified by the Council on Naturopathic Medical Education. I had no idea what naturopathy was, but LoBisco quickly explained. "Naturopathy taps into the body's ability to heal itself. It avoids drugs and surgery, and instead of treating symptoms, gets at the root cause of an illness."

Coming from my perspective as a geneticist, I was perplexed. Wasn't the root cause of an inherited illness a mutation, a change in DNA that only gene therapy can fix? Don't the tools of naturopathy—eating whole unprocessed foods; taking vitamin, mineral, and food supplements; practicing yoga, meditation, and exercise therapy—look to counter a symptom that a genetic glitch sets into motion? My confusion grew when I later read on the website for the NIH's National Center for Complementary and Alternative Medicine, "Practitioners seek to identify and treat the causes of a disease or condition, rather than its symptoms." It was a little like the new cystic fibrosis drugs that are claimed to treat the disease at its source, but which actually target the protein, offering a treatment that is daily, not forever. Semantics aside, the field began with Benedict Lust, who was treated by a natural healer in Germany and brought the approach to the United States in the early twentieth century, christening it naturopathy. This holistic, wellness-centered, noninvasive movement spread for three decades, until antibiotics and other advances in conventional medicine began to outshine it. Naturopathy is popular again today, not as an alternative method of healing, but as a complement to modern medicine—as it is for Hannah.

LoBisco listened intently as Lori's complicated story spilled out in torrents of jargon, typing notes and occasionally googling when the name of a molecule surfaced from the narrative stream. The conversation wandered as Lori flitted from science-speak to mommy-speak to spelling the multisyllabic words so that Hannah, who can read and understand at a level much higher than the norm for her age, wouldn't catch on. As the doctor guided Lori through a detailed family medical history, a larger picture emerged.

Lori described the regimen of alternating dietary oils—wheat germ, evening primrose, black currant seed, sesame, and fish—that Hannah had recently begun. These would support her cell membranes, according to the chiropractor who recommended them. She talked about a food supplement, Neurotrophin, that she chopped up and sprinkled onto Hannah's food. The stuff is mashed pig brains. The language on the website that sells it is technical and vague at the same time, a curious mix of language that would seem accurate and impressive to anyone who isn't a biologist. It is sneaky, too. The name of the pig brain extract emulates "neurotrophin," a real type of nurturing molecule in the nervous system. With a capital letter and a trademark symbol, it *sounds* like a drug, but it isn't, because "food supplements" do not need FDA approval to be marketed.

As LoBisco googled away, asking questions and offering ideas and suggestions, she and I gradually realized that our areas of expertise are complementary, not in conflict. She talked about probiotics and organic foods, I talked about genes and proteins. Meanwhile, Lori segued smoothly from detailing how a missing protein hugely inflates the axons of Hannah's nerve cells to a vivid description of the length, consistency, hue, and odor of her daughter's most recent poop. Hannah sat quietly in the corner, reading, but as the two-hour mark approached, she grew cranky. She was, after all, a six-year-old, although at times she has a pensive look in her eyes that seems older. LoBisco still had a few pages to complete in the history, so she turned her computer around, and Hannah soon had Barbie on YouTube prancing about on the screen.

When Hannah stumbled upon a restricted website with naked people, it was time to go. LoBisco ended the session with recommendations to eat organic foods and to try eliminating milk to address Hannah's stomach pain. She would need time, though, to learn more about Hannah's disease, so that she could determine whether naturopathy could help.

It's understandable that LoBisco didn't know anything about giant axonal neuropathy—GAN. Most MDs don't, either. For there are only forty-five recognized cases in the world, and Hannah Sames is the youngest.

Few disease names have as clear a meaning as GAN—*a nerve disease of giant axons.*

Neurobiology 101: An axon juts from the central cell body of a nerve cell, or neuron, like a tadpole's tail. On the other side of the cell body bursts an explosion of branchlike dendrites. Looked at lengthwise, a neuron is a little like an out-of-proportion upper limb—the dendrites represent the fingers, the axon the arm, and the cell body the hand itself. If the cell body were the size of an apple, then the brushy dendrites would bloom outward several feet. The axon is considerably longer. Dendrites receive messages and axons send them.

A cross section of healthy white matter nerve tissue looks like a sheet of tiny dots when looked at head-on. These dots are aligned axons. In GAN, some of the dots are huge circles, grossly swollen, but they aren't empty. Hannah's axons are stuffed with strawlike fibers called neurofilaments. How the axons got that way remains a mystery, despite knowing the root genetic cause.

Sporadic descriptions of what might have been GAN dot the medical literature from the 1950s, when it was dubbed a vague "polyneuropathy." In 1971, researchers at the University of California, San Francisco, discovered the giant axons and named the disease, and noted a curious symptom—extremely kinky, fair hair. Their subject was a six-year-old with a medical history eerily like Hannah's. She met her early developmental milestones, but by the time she was three, neighbors began to comment on her clumsy gait. The parents were so accustomed to how their little girl walked that they didn't realize her awkwardness could be an indicator of a larger problem. A physician diagnosed "rheumatism" and advised that the girl take aspirin for a few weeks. If that didn't help, he'd start a workup for muscular dystrophy.

The girl became weaker. In another six months, her legs were unsteady and her feet oddly turned out. She waddled. Then it became difficult for her to hold things. Her speech became slightly slurred, but she still did well in school. Then she could no longer rise from sitting. This was a sign of muscular dystrophy, but a muscle biopsy at age five showed normal muscle structure. Plus, there was no family history of any muscular condition.

By the time the child visited the physicians in San Francisco, her legs were more splayed, and she walked hesitantly, on the outer edges of her feet. Her arms and legs were weak and wasting away, but the cords of muscle didn't jump and lurch as they do in amyotrophic lateral sclerosis (ALS, also known as Lou Gehrig's disease), where they look like snakes slithering beneath the skin. She could no longer move her toes or flex her feet, and she was beginning to have difficulty moving her fingers. Sensation had faded away. Viewed under a microscope, her kinky hair was unlike the corkscrew curls of other inherited conditions, such as Menkes disease.

Whatever their young patient had, the doctors knew it hadn't been described before. When they sampled a bit of nerve tissue from near her ankle and looked at it in cross section, they were shocked to see what resembled Swiss cheese: axons of varying sizes, some swollen to "enormous proportions." The researchers described "massive whorls" and "tightly woven . . . parallel skeins" of neurofilaments that had squished other cell parts to the sides. As a result, the muscles in the child's limbs weren't receiving signals to contract.

Three years later, another case appeared in the medical literature from a different group of researchers, who urged doctors to suspect GAN "in a patient with tightly curly, pale scalp hair, unlike that of his parents." By 1987, only twenty more cases had been reported. That year, researchers from the University of Vermont described a twelve-year-old girl who was sicker than the original six-year-old and Hannah. She had problems crawling, couldn't walk until age two, and was crashing into things by age three. But she also had heart problems, scoliosis, incontinence, early puberty, and her eyes flitted back and forth. When neurologists reported families with two affected children in 1998, it became clear that the disease was inherited, and the fact that these two dozen patients recognized so far were children was ominous. However, also nestled into the medical literature were reports of adults with symptoms similar to those of inherited GAN from families without other cases. For them, the cause might be exposure to certain chemicals. "Glue sniffer's neuropathy" from solvents and "leather cement poisoning" from acrylamides in the shoe industry are but two causes of environmentally induced GAN.

By 2000, a research team from several European nations had found enough affected families to search for a DNA sequence that the children with GAN had in common. This is a standard experimental approach in genetics, dating back to the first genetic map circa 1984—meticulously compare DNA sequences among people with an illness, and with those of healthy individuals, perhaps the patients' unaffected siblings. Then search for mutations seen only in the sick kids. The next step is to determine if the function of the specified protein makes sense, given the symptoms, and this stage may require more experiments and/or lucky or creative thinking.

For GAN, such a gene search led to chromosome 16, to a DNA sequence that provides the instructions for cells to make a protein colorfully named gigaxonin. Hannah is missing part of that gene in each of her two chromosome 16s. She got one copy from Lori, and one from her father, Matt. If it was normal, Hannah's gigaxonin would keep the protein filaments in her motor neurons neatly aligned, so that the cells could effectively reach from her spinal cord down to her toes. Instead, some of her motor neurons are a bloated mess.

Chaos reigns in the footlong nerves that reach down Hannah's legs. The problem lies in the scaffolding that gives a cell its characteristic shape. The cytoskeleton—literally "cell skeleton"—has girders of three sizes. The smallest and the largest girders are each made of one kind of protein, but the medium-sized girders, the intermediate filaments, include at least five types of proteins in nerve cells. In hair, they have a different ingredient list, and it is the same recipe as that for nails, horns, and hooves. Hannah's intermediate filaments are in great disarray. Early on, they only kinked her hair. Now, neurofilaments are filling the axons of her nerve cells, slowly shutting down their ability to communicate with her muscles.

An axon looks like a tail, but it doesn't swish to move the nerve cell of which it is a part. Instead, an axon enables the neuron to transport messenger (neurotransmitter) molecules from where they are made in the nucleus to the outer reaches of the cell, the very tip of the axon, from which the molecules are released into the tiny gap, or synapse, between the neuron and another neuron or a muscle cell. The neurotransmitter molecules bind to these "postsynaptic" cells, triggering electrical changes that spread the message. Such neurotransmission is a complex choreography that goes awry in many diseases.

The neuroscientist Tony Brown, PhD, heads a lab at the Center for Molecular Neurobiology at Ohio State University that studies normal neurofilaments, in hopes of discovering what goes wrong in diseases such as ALS and GAN. He describes a healthy axon: "Everything in the axon is moving. It's a massive highway with thousands of pieces of cargo, moving at different rates, powered by different motors, regulated in different ways. Some of the cargo has a destination, like being shipped to the axon's end to deliver a message to a muscle cell or another neuron. Other cargo move more slowly and intermittently and not to a particular destination, just replenishing things. Neurofilaments are probably these. Such a complex highway is tremendously vulnerable. Perturbing the momentum over time results in logjams."

Hannah's neurofilaments are horrifically deranged, both too many and too large. Developing a treatment starts with understanding how the abnormal gigaxonin protein affects axons of the motor neurons and causes the symptoms. This is where "animal models" come in, other species that naturally have a disease of humans, or do so because we alter them to have certain human genes. GAN mice are helping researchers to understand what goes wrong in children with GAN. While the rodents aren't nearly as sick as the children, their axons are far from normal.

Yanmin Yang, MD, PhD, at Stanford University leads one of three groups of researchers worldwide who have made GAN mice. The investigators start with genetically identical mice and then remove, or "knock out," the GAN gene in only some of them. Like Hannah, Yang's altered mice lack enough of the gigaxonin gene to hamper the ability to move. These mice, by six to ten months of age, are not as active as their healthy littermates and have a barely noticeable shift in gait. Gradually, their back legs splay as they weaken and the rodents may become spastic and develop seizures. The GAN mice also have curly red hair, baldish spots of bent hairs, and no whiskers, and they drag their rear ends. By a year of age, muscle wasting and weight loss worsen, and the mutant mice stumble about their cages. However, most survive until twenty months, which is about the equivalent of a person living into his or her sixties. Looked at under a microscope, the neurons of the sick mice have big axons. They have fewer than their human counterparts who have GAN, but the axons are stuffed tightly with neurofilaments and tiny bubbles stuck in the logjam.

Understanding how a disease develops entails understanding what the implicated gene normally does in affected cells. To do that, Dr. Yang looked at normal gigaxonin protein in brain cells from healthy mice. She found that one end forms a propeller that touches another type of protein that in turn controls how the nerve cell handles the normal buildup of protein debris. Healthy gigaxonin regulates levels of that garbage-control protein, which maintains the sleek structures of the cell's scaffolding. Without gigaxonin, the cell's garbage control lifts, and the scaffolding overgrows, bulging the axon. (When Lori mentioned this finding to Dr. LoBisco, the naturopath, she googled for food supplements that might affect the garbage-control protein. So she *was* looking at a "cause of the illness," albeit one removed from the gene itself, but recommending a supplement based on an educated guess, not the findings of a controlled clinical trial.)

GAN mice might have an easier time than GAN humans for a practical reason—the animals distribute their weight onto four limbs, while humans, perhaps, stress their two legs enough to cause severe symptoms. But the really good news about Yang's GAN mice is that they live near-normal life spans. Might children, too?

On March 5, 2004, when Lori and Matt first glimpsed their newborn daughter Hannah's kinky ringlets, they were both delighted and puzzled. Their other daughters, Madison, five, and Reagan, two, had stick-straight hair, as do Lori and Matt. When the birthing goop had dried, Hannah's curls were odder still, weirdly dull, like the "before" photograph in an ad for a hair conditioner.

For a long time, Hannah seemed normal. She smiled, sat, crawled, and hauled herself upright on schedule. The first sign that something might be wrong wasn't even very alarming. Hannah's Grammy Judy noticed that the toddler was taking halting, hesitant footsteps. The grandmother thought it was probably just the uncertainty of going from crawling to walking. But instead of becoming more sure of herself and coordinated, Hannah slowly grew clumsier, as the strength ebbed from her legs. Lori took her to an orthopedist and a podiatrist, who reassured her that Hannah would be fine. But already, feathery filaments of protein were filling the long axons in her legs, blocking the nerve impulses to her muscles.

Hannah grew into a charming little girl who loved to sing and dance, dress and undress her Barbie dolls, and play outdoors with her sisters, "a beaming light of love," says her mother. But by Hannah's third birthday, Lori and Matt suspected something was seriously wrong. Their daughter didn't walk normally, she tottered, her lower legs bowing out. Her pediatrician gave her a rigorous physical exam and agreed her walking was off, but, like the two physicians before, told the anxious parents, "That's just how Hannah walks." She'd likely outgrow it. Unconvinced, Lori took Hannah to another orthopedist. Diagnosis: "Just let her live her life." She would be fine. The unspoken diagnosis: helicopter mom syndrome.

Lori and Matt remained certain that something was wrong, as only parents who watch a child 24/7 can. Then came the break. "My sister showed cell-phone video of Hannah walking to a physical therapist she works with who thought Hannah's gait was like that of a child with muscular dystrophy," Lori says. Since muscular dystrophy is an inherited muscle disease, the local pediatrician then referred Hannah to a neurologist and a geneticist, and six months of testing for various nerve and muscle diseases ensued. The results: all normal.

But Hannah wasn't normal at all.

Lori will never forget the moment when she learned what her little girl had. The answer didn't come from a DNA test or a medical scan, but from an astute pediatric neurologist. "He took out a huge textbook and showed us a photo of a skinny little boy with kinky hair, a high forehead, and braces that went just below the knee—he looked *exactly* like Hannah. And he had GAN."

Parents of very sick children remember diagnosis dates in the way that everyone remembers birthdays or dates of death. Hannah Sames was diagnosed on March 24, 2008. The next day the family headed down to a New York City children's hospital for three days of tests and exams to confirm the diagnosis. The workup was more intense than assessing gait, including biopsies and spinal taps. And the neurologist with the huge textbook turned out to be right.

Back home, Lori and Matt met with a genetic counselor, and that is when the devastation really set in. A genetic counselor is a health care professional who explains how genetic illnesses happen, from the molecular particulars to the big picture. Lori gets a faraway look in her eyes as she recalls that session and dully recites what they learned.

"Matt and I are each carriers of GAN, and we passed the disease to Hannah. Each of our two other daughters has a two-in-three chance of being a carrier. GAN is a rare 'orphan genetic disorder' for which there is no cure, no treatment, no clinical trial, and no ongoing research."

"So you are telling us this is a death sentence?" Lori recalls asking the genetic counselor.

"Yes," she answered quietly.

The disease would progress slowly, the counselor continued. Hannah's legs would gradually weaken. By first grade she'd likely need a walker in addition to her ankle supports, and soon after, a wheelchair. She might lose her sight and hearing, and eventually be bedridden.

Matt and Lori walked around like zombies for a few days. Then, like the Salzman sisters who dropped everything to form Stop ALD, and Michaela and Augusto Odone before them who developed Lorenzo's oil, the Sameses became activists, fund-raisers, and amateur scientists. For their disease—GAN—had no Michael J. Fox, no Christopher Reeve, not even a Lorenzo to fortuitously interest a filmmaker. What pharmaceutical company would be interested in treating a disease that affects a few dozen individuals?

Unlike the Salzmans, Matt and Lori had no connections to the scientific community or the biotech or pharmaceutical industries. Nor was a gene therapy already in the works, as it was for Corey and Ashi. Matt and Lori would have to attack the mysterious illness alone, and from scratch.

11

LORI

On a spectacular Sunday morning in September 2010, seven-hundred-plus people gathered on the streets of a suburb north of Albany, New York. The winding roads of the neat development blended into a montage of large homes, all beautifully landscaped with nary a leaf out of place. Intricate multicolored chalked messages—"Go Hannah, You Rock!"—colored the driveways. Every twenty yards or so a pink "Hannah's Helper" sign with a child's name on it poked up from a perfectly manicured lawn.

The five-kilometer Run for Life was to benefit Hannah's Hope Fund, the not-for-profit organization that Lori and Matt Sames started almost as soon as their daughter was diagnosed with giant axonal neuropathy. The hour before the race was quasi-organized bedlam. Small groups of runners sporting matching tees gathered, from a pediatric dentistry practice to a local sandwich shop to a law firm. Knots of teens from the nearby middle school dominated the scene, and young children waited patiently for their turns in the huge Hamster Ball or Mr. Bouncity Bounce. Bruce Springsteen's "Born to Run" blasted through the air. Pots of mums and baskets of apples heralded the arrival of autumn in this community at the foothills of the Adirondacks.

I looked around in awe at the outpouring of love for Hannah and her family. This is what Corey's parents were spared, the relentless and exhausting fund-raising necessary to plan and execute a clinical trial to test a treatment for a disease that no one has ever heard of.

On the lawn in front of the registration tent, a small girl with frizzy red hair and a #1 racing tag pinned to her pink T-shirt pushed a pink stroller containing Stella, her cousin's Yorkie. A cameraman from a local TV station bent down to film the little girl, who was unaware that the stroller was functioning as a walker. Channel 6, WRBG Schenectady, is following Hannah through her entire journey toward and during gene and other therapies.

As race time neared, the family assembled on the lawn near Hannah and the crowd sorted itself out at the starting line by running ability. Lori grabbed a flag as Matt hoisted Hannah up onto his significant shoulders. Her older sisters, Reagan and Madison, stood in front of their dad; they all waved at the runners. Lori, in ponytail and jeans, took the microphone, and after a few thank-yous, characteristically got right down to business. "Here's the scientific update. We're awaiting efficacy endpoints within the next eight weeks. So at next year's race, we'll be able to say we're two months, or four months, from treatment. *You guys have made this possible!*" she yelled.

Madison and Reagan, small versions of their mom, sweetly sang "The Star-Spangled Banner," as Hannah tried to join in from above. Then the emcee, Hannah's aunt, grabbed the mic, said "Little Miss Hannah will now start the race," and passed it up to the guest of honor. *"On your mark! Get set!"* Hannah shrieked, and then her *"Go!"* was drowned out by a horn. The runners surged forward.

Getting seven hundred people out early on a Sunday to run through a small neighborhood on an activity-packed weekend in the dying days of summer was quite a feat. But it was nothing compared with what had happened just three weeks earlier.

Hannah's Hope Fund for GAN was born in Lori and Matt's basement. Their business backgrounds helped. Lori had been a project leader for computer software installation for health care organizations. Matt runs the family business, a Pet Lodge that provides boarding, sitting, doggie day care, and kitty camps. Right after the diagnosis, Matt and Lori got to work.

The first step was to assemble a team of scientists who might be interested in trying to develop a treatment or cure for GAN. Lori and Matt would get them all together at a conference they would sponsor through fund-raising. "The basement was like a dungeon. We had an old desktop computer, lots of Sharpies, and a huge white-board that we called our war board. Matt's very visual, so we'd look things up on the computer and he'd start putting up names and numbers," Lori recalls.

Soon their online research led to Jude Samulski, director of the Gene Therapy Center at the University of North Carolina at Chapel Hill and 2011 president of the American Society of Gene and Cell Therapy. "I left a voice mail, and he called back right away from a meeting in Europe. He focuses on muscular dystrophy, but he had a young investigator who might be very interested. He'd send him to our symposium to learn more about GAN to see if they could help," Lori says. The UNC group was developing adeno-associated viruses (AAVs) as vectors, the type that would be used on Corey, and the scientist Samulski had in mind was Steve Gray, a new PhD. Barely out of his twenties at the time, Gray is still so boyish in appearance that Lori calls him Doogie Howser, after the precocious sixteen-year-old doctor at the center of the 1990s TV show of the same name. He still gets carded when he orders a beer.

While putting together their scientific dream team, Lori and Matt had to deal with what to tell their other two young daughters. Madison, then eight, knew something was up. Why were her parents suddenly on the phone all the time, or the computer? Why didn't they sleep? Why were they always in the basement scribbling on a whiteboard and getting so excited?

With help from Lori's sister, Matt and Lori got a website up and running, with a video in which they tearfully described what Hannah will face. "Madison heard at school that kids had seen the video, and they said, 'Your mom is crying!' So I explained that we need to get money to pay doctors to help Hannah walk better," Lori says. A few weeks later, Lori walked into a father-daughter fund-raiser in Saratoga to find Reagan huddled on a bench, crying. "Is Hannah going to die?" the little girl asked. Her best friend had seen the video.

Lori and Matt had discussed at great length what to tell the girls, delaying the inevitable conversation as long as possible, hoping that perhaps Madison and Reagan would see only the current limitation of Hannah's walking ability. They didn't want to lie but at the same time couldn't bring themselves to paint the whole grim picture for a six- and an eight-year-old. But at the same time, they had to get the word out to raise funds and awareness. The visibility of the website finally forced them to clearly communicate the facts to their daughters. "So I told Reagan that Hannah could die, but that probably wouldn't happen, and instead of thinking about that, it's better to try to get her to walk better," Lori recalls, tearing up. They explained to Madison and Reagan that Hannah didn't know what all the attention was about. She just knew she couldn't walk well.

As if reading the medical literature, setting up a scientific conference, raising three young children, and entertaining cats and dogs wasn't enough, the Sameses also threw themselves into fund-raising. They quickly learned how much a clinical trial, and all that led up to it, would cost: millions of dollars. They'd have to fund the preclinical studies on an animal model, choose among therapeutic avenues such as gene or stem cell therapy, find researchers who would do the experiments and perhaps fund the work if government or other grants wouldn't cover expenses, then help to develop a clinical trial protocol and, finally, get families to sign up their children for an experimental therapy that had absolutely no guarantee of helping, and that could possibly even cause harm. The road ahead was overwhelming. Hannah's Hope Fund adopted Margaret Mead's mantra, "Never doubt that a small group of thoughtful, committed citizens can change the world. Indeed, it is the only thing that ever has."

They held barbecues. Golf tournaments. Balls. Silent auctions. Hockey games. Dance marathons. Vacation giveaways. They sold T-shirts and holiday cards and hoodies emblazoned with a cartoon icon of Hannah. Lori and Matt talked about Hannah and GAN wherever and whenever anyone would listen to them. In the capital region of New York, it became impossible *not* to know of Hannah and her story. I'd first heard about her shortly after her diagnosis, from a front-page article in the *Daily Gazette*, Schenectady's hometown newspaper, and I later wrote about her giant, stuffed axons in my human genetics textbook, just as I'd learned about Corey in the local paper and put him in my textbook. But I didn't meet Little Miss Hannah until a mild Saturday afternoon in January 2010, when she perched next to me on a bar stool at a local music venue, a rock concert in her honor about to get under way.

It was the day after the celebrity-studded Haiti earthquake relief concert on TV, and people might have been donated out, but the place was packed. Hannah told me about her dog Ginger and I told her about my snoring cats. "Cats snore?" she giggled. Her wiry hair stood strangely still as her head shook, due to a microscopic peculiarity of GAN—the diameters of the hair shafts are uneven, which somehow disrupts the sleekness of hairs, just as the mutation disrupts the sleekness of nerve axons, so that the curls stiffen. Some of her hair was caught up in two ponytails, and with her impish grin, she looked like a pixie. When the music began, she clamped her hands over her ears and grimaced, but as a series of well-wishers took turns snatching her up in hugs and whisking her around the room to the beat of the music, she smiled and laughed.

Lori slid onto the abandoned bar stool to chat, shouting above the throbbing music. I'd already interviewed her for my textbook, so I was prepared for her unique conversational style, intense and rapid, ricocheting from mom mode to spouting the details of the viruses that would one day carry healing genes into the spinal cords of Hannah and eight other children. But animal experiments would come first. That day Lori was excited about testing the virus on farm pigs at Emory University, and the fact that Hannah's Hope Fund had already raised more than $1.2 million. "We're making our miracle happen!" But Lori and Matt knew that millions more would be needed to fund the leap from animal studies to executing a clinical trial, especially on young people who, like Jesse Gelsinger, were not all that sick.

The fund-raiser that placed Hannah's Hope Fund securely on the path to gene therapy was the Pepsi Refresh Grant competition. Every month Pepsi gives away $1.3 million to fund worthy projects. Anyone can vote, up to three times a day, by text or online. Topics range widely, although diseases, especially of children, are favorites, as are programs to combat poverty and help the survivors of natural disasters. Many projects are pet-centered, such as building parrot shelters, opening a food pantry for dogs, and replacing rusted kennels. Each month, ten grants are awarded for each of the $5,000, $25,000, and $50,000 categories, but only two succeed in winning $250,000. Since Lori Sames is incapable of thinking small, Hannah's Hope Fund went for the big prize.

The voting began August 1. On that day, 1,224 nonprofits eyed the $250,000 prizes. By the end of the day, HHF was in the 400s. A steady rise started as Lori, Matt, and their girls convinced everyone they had ever met and many total strangers to vote, three times a day. The nine other GAN families in the United States did the same in their communities, and the voting went viral. By mid-August, Hannah's Hope Fund slowly crept upward—10—9—8—sometimes jumping a few points in a day. The local media followed the story, as Lori became a fixture at the nearby Saratoga Racetrack, where she begged hundreds of complete strangers, from socialites to guys drinking beer around trash cans, to vote for Hannah's Hope Fund.

By the third week of August, HHF stalled at number 5 for a while, reached number 4, and then slid back to 5. This sent Lori into warp drive.

The last week of August was harrowing. HHF had reached number 3. The International Rett Syndrome Foundation held the number 1 slot. Julia Roberts was their spokesperson exactly when her film *Eat, Pray, Love* opened, so it looked hopeless for Hannah.

Project number 2 was Cure Juvenile Myositis. Those two would be hard to beat, but Lori, Matt, and their helpers weren't daunted.

With two days to go, as if by a miracle, Hannah leaped to number 1. At 10:45 the morning of the last day, Lori checked the Pepsi Refresh website. Hannah was still holding strong, but Lori wasn't going to take any chances. In 95-degree heat, she and a girlfriend stood in the traffic outside an atomic power plant when the 3:30 p.m. shift got off, stopping drivers. When the traffic diminished, they walked up the road a quarter mile to the General Electric Global Research Center, which let out at 5:00.

At 5:30, Lori got a call from a friend whose organization was competing for one of the smaller Pepsi grants. They'd just slipped from 7 to 9, the friend warned, so things could change in a heartbeat for Hannah, too. Although HHF was still a comfortable 1, Lori panicked, and began frantically texting and e-mailing the immediate world.

At 9:30 p.m., August 31, Hannah's Hope Fund dropped from 2 to 3, even as we were all voting continuously, the three-times-a-day rule long abandoned. Later, Lori realized the dip was due to the West Coast support for the other charities kicking in at the end of the day, especially that of the ASPCA, whose popular rusted kennel project threatened to bump Hannah. (It was ironic, considering the family business of pet care.)

"So I freaked out. I called Matt, who knew someone who knew someone who was chief of staff at the University at Albany, who got the students to go door to door. I ran through the supermarkets. I approached people coming out of movie theaters. A sheriff was parked outside, so I asked him to put it out on the radio, and I asked local cops and state cops, too. A GAN family in Georgia got truckers to radio it," Lori told me breathlessly on the phone the next day, between TV interviews.

Meanwhile, every local TV station was carrying the story, with a continuous scroll on the bottom of the screen urging viewers to vote. Corey and his parents voted over and over too, from their home in nearby Hadley. The cities of Albany, Schenectady, Troy, and Saratoga were whipped into a shared frenzy of texting and online voting that crashed Pepsi's server. But none of us knew that. All we could see on our screens as the clock ticked down in the final moments of August 31, 2010, was that Little Miss Hannah was stalled at number 3. I cried at midnight, and when I finally fell asleep, had nightmares of the morning after the Bush/Gore tied election of 2004. I could hardly imagine how Lori and Matt felt. It turned out, though, that Lori had never given up hope.

"Last night when I went to bed, after taking a sleeping pill, I had a little twinge in my gut. I thought we could wake up to good news. I knew texting was not in real time, and that it was backing up. I opened my eyes at 6:16 a.m. and went to my computer. And I saw 'Finalist' in the right corner!"

Confusion reigned for a few hours, though, because the 3 still stood beside Hannah's Hope Fund on the website, but Senator Kirsten Gillebrand's office—her former territory includes Hannah's community—confirmed the win with Pepsi and called Matt. He was in the winner's circle with his family up at the Saratoga track, where a race had just been run in Hannah's honor. Lori got the confirmatory call from Pepsi at about the same time, 3:30.

For Steve Gray, who would be designing the gene therapy for GAN, the win was bittersweet, for he'd been involved with Rett syndrome too. It causes autism, intellectual disability, breathing problems, and sometimes early death, and affects only girls because it is lethal before birth in boys. He blogged, "Part of my research with the International Rett Syndrome Foundation led me to solve a critical problem in our development of a GAN treatment. While I celebrated HHF's victory, I was also sad that the IRSF couldn't get the money too. If you can keep your enthusiasm going another month, please throw your support behind them." We all did, and in September 2010, the International Rett Syndrome Foundation won its $250,000 Pepsi Refresh grant.

The only thing that stopped Lori in her tracks during those crazy days in late August 2010 was when she heard from GAN families with affected children in their early twenties, such as a distraught sister who begged for Hannah's not-yet-tested treatment for her twenty-three-year-old brother. "I can't bear the magnitude of this solely on my shoulders," Lori e-mailed her research team. The neurosurgeon at the University of North Carolina who will work on Hannah and the others assured her that she can't be responsible for everyone dying of the disease. But the other parents are never far from Lori's thoughts.

Lori and Matt learned a great deal from other parents who had faced untreatable inherited diseases in their children. Even the famous are not immune from genetic fate, but it certainly helps, from raising funds to getting the attention of legislators, to have a celebrity personally connected to a disease. Even if the child dies, science can benefit, and help others. This was the case for Hunter Kelly, the son of the former NFL quarterback Jim Kelly and his wife, Jill.

Hunter was born in 1997. He cried constantly and had difficulty nursing, and then his arms and legs stopped moving. His back arched and limbs jerked in the purposeless movements that indicate severe brain damage. When Hunter was nine months old, doctors finally had a name for his condition—Krabbe disease, or globoid-cell leukodystrophy in the new lingo of biologically meaningful, if unpronounceable, descriptors. Like the disease that struck Lorenzo Odone and Oliver Lapin, Krabbe is a demyelinating condition, but Krabbe starts its relentless march much earlier. Newborn screening would have picked it up. After Hunter was diagnosed his remaining motor skills quickly ebbed and his mental development ceased. He needed a feeding tube and lost his hearing and vision. Although Hunter's father's fame enabled his grandmother to start Hunter's Hope to raise funds, the boy lost his battle at eight years old.

President Bush signed the Newborn Screening Saves Lives Act in April 2008, which would institute a standard slate of such tests, but the funding ($44.5 million) has not come through, and so the number of tests performed on the blood taken from the heels of newborns varies from state to state. At the present time, only New

York screens newborns for Krabbe disease. But Hunter's Hope has raised awareness and galvanized other families to try to get legislation for more extensive newborn screening passed in their states. The value of newborn screening for this particular disease is that if the mutant gene or a telltale altered metabolite is detected before symptoms begin, an umbilical cord stem cell transplant can prevent symptoms, as it did for Spencer, who is enjoying a normal boyhood after being treated for the ALD that claimed his cousin Oliver. Would newborn screening for GAN someday be teamed with a treatment that could prevent disease progression?

Lori and Matt were deeply moved by the various parent-run not-for-profit organizations they learned about, as they were by the researchers they encountered, eager to explore a rare disease. But at the same time, they learned that the field of rare diseases was sometimes contaminated by individuals seeking to profit. The situation was disturbing. If children's tissues were to be used to develop tests and treatments, donated before or after their deaths, then protection of those precious resources was necessary. The Sameses were drawn into this world by a woman who had become legendary in the rare-disease community, Sharon Terry. She is nearly single-handedly responsible for activist-parents claiming ownership of intellectual property resulting from their children's contributions of their abnormal cells and tissues. In addition, her personal story showed Lori and Matt the mountain that they were about to scale.

Sharon and Patrick Terry hadn't expected anything dire when seven-year-old Elizabeth developed a rash on her neck. The beige bumps looked like dirt, but they wouldn't go away with scrubbing. Just before Christmas in 1994, Elizabeth's pediatrician examined the rash and told Sharon and Patrick not to worry—so they didn't. When the strange bumps appeared on five-year-old Ian, too, the concerned parents took the children to a dermatologist. The diagnosis: pseudoxanthoma elasticum, or PXE.

Like Corey's disease and Hannah's, PXE is inherited from unaffected carriers, and typically comes as a surprise. Calcium is deposited in the skin, the eyes, the linings of blood vessels, and in the digestive system. Sometimes it gets no worse than a scruffy-looking neck or pale bumps under the arms and behind the knees. Other times it clogs the circulatory system, like atherosclerosis in a much older person. Because the artery clogging could be deadly, Sharon and Patrick knew they had to take action.

Like the Odones, Sharon and Patrick hit the medical library, using the one at the University of Massachusetts in Worcester, a short drive from their home in a Boston suburb. Even though PXE had no treatment, it wasn't exceedingly rare, with 11,000 patients in the United States alone. If the Terrys could collect DNA and give it to researchers who could find unique sequences that PXE patients share, it could lead to a drug target—a specific molecule that a specific drug would bind and destroy or deactivate. So the couple applied their strengths to their terrifying new reality. Sharon was a geology and theology major who was once a college chaplain; Patrick was an engineer who'd worked at some of the biotech companies that cluster around Boston. Like Lori and Matt, they rapidly mastered the art of fund-raising.

The Terrys founded PXE International, which had the goal of linking researchers and providing DNA to help them identify the mutant gene behind PXE. The not-for-profit organization coordinated patient services, established a blood and tissue bank and an extensive database, and now oversees thirty-three labs and fifty-two offices throughout the world. Sharon Terry was a coauthor of the two technical papers unveiling the PXE gene in 2002. More important, she was a co-inventor on the patent for the PXE gene, with all rights assigned to PXE International so that the organization could fund testing. It would be maddening for the children who had provided tissue that made a diagnostic test possible to be charged for taking those tests.

I told Sharon Terry, whom I met at a genetics conference, all about Hannah's Hope Fund's remarkable progress, and she offered advice. "Lori and Matt, you have already done great things—to bring together researchers on a rare disease is half the battle. Now, get bold. Figure out which other conditions share any commonality with giant axonal neuropathy. Meet with the people behind any of those projects. Join forces. While you wait for the research to catch up to your dreams, get the affected individuals ready for clinical trials. Collect clinical information in a registry, collect blood and tissue in a biobank. Look for correlations and trends in the data. Understand what the community needs and the natural history of the disease. Figure out endpoints to use in drug development."

Lori quickly heeded Sharon Terry's advice in learning from other diseases, both to better understand GAN and to have a ready answer to researchers or funding agencies that might ask, "Why should we care about such a rare disease?" Lori chose as her model spinal muscular atrophy, or SMA. Like GAN, SMA is autosomal recessive—it affects both sexes and is inherited from carrier parents. The recent saga of SMA activism shows how a disease can go from obscurity to the very brink of gene therapy.

Both GAN and SMA affect motor neurons, which signal muscles to move. (In contrast, the muscular dystrophies—such as Duchenne and Becker muscular dystrophies—remove proteins from muscle cell membranes that otherwise enable them to withstand the force of contraction.) In GAN, motor neurons don't work because their axons bulge. In SMA, the motor neurons shrink away from the muscles. It is heartbreaking to watch. A baby with SMA who has learned to sit, crawl, stand, and walk regresses until he or she can't move at all. SMA is called "baby ALS," and like that disease, the mind stays alert as the body fails. Three subtypes are commonly recognized, based on severity.

Like GAN, SMA has no Michael J. Fox, Julia Roberts, or sports hero to plead its case in Congress for funding. But unlike GAN, SMA has numbers: 25,000 people are affected in the United States, and about 40 percent of them are adults. One in 6,000 newborns has SMA, and 1 in 35 people is a carrier, amounting to about 7 million people who could pass the disease to a child if their partner is also a carrier. "Spinal muscular atrophy kills more babies than any other genetic disease," states the FightSMA website. It is almost as prevalent as cystic fibrosis, and more common than sickle cell disease, muscular dystrophy, or ALS. Yet few people have heard of it, and funding is sparse. At the same time, enough is known about the disease—its gene, protein, and mutations—that it is an ideal candidate for gene or other therapy. And mastering gene delivery to the motor neurons could pave the way toward a much more common condition that cuts off these cells—spinal cord injury.

In 2003, when fifty scientific luminaries sent a letter to the director of the NIH saying that all it would take to get a treatment for SMA within five years was funding, the purses opened. Late that year, the National Institute of Neurological Disease and Stroke announced a new translational research ("bench to bedside") initiative, with SMA the first targeted disease. Legislation also singled out the disease. H.R. 2149 (a bill before the House of Representatives for the 111th Congress) seeks to "authorize the Secretary of Health and Human Services to conduct activities to rapidly advance treatments for spinal muscular atrophy, neuromuscular disease, and other pediatric diseases, and for other purposes." Within the broad mandate, the details of the bill were written for SMA.

In 2010, gene therapy worked for SMA—in mice. Two independent research groups delivered the healthy human SMA gene, aboard adeno-associated virus into facial veins of mice that have a version of the disease. With a single injection on the first day of life the animals could move, and they lived more than 250 days, compared with fifteen days for those with the illness. The shot was less effective if given on day five, and didn't work at all if given on day ten. The falling efficacy with time revealed the treatment window. The cured mice sent ripples of excitement through the SMA community.

I sent Lori a news release announcing the very encouraging SMA mouse work, and just reading the words "SMA is a devastating motor neuron disease which affects children" was enough to convince her that she had to talk to the leaders of the SMA community, who would be at the upcoming annual meeting of the American Society of Gene and Cell Therapy in Washington, D.C. If Hannah's

Hope could join one of the SMA organizations, perhaps the GAN community could tap into some of the exposure and clout of the farther-along SMA effort. And so at the meeting, in May 2010, Lori and Steve Gray sat at a Washington, D.C., hotel bar with Christian Lorson, science director for FightSMA and a professor at the University of Missouri, and Karen Chen, research director of the SMA Foundation. Lori breathlessly recounted ongoing experiments in a stream-of-consciousness burst of jargon that nevertheless made sense to Chris and Karen.

"Nick Boulis at Emory is treating nine farm pigs. They're getting AAV9 carrying GFP, but the buffer killed some of them, and we're waiting to hear about the last five, and next we'll hook up the gigaxonin gene, and oh, it is so expensive to do large animal studies, farm pigs aren't so bad, but primates . . ." Lori paused and shook her head. "Can't we all just *share* the monkey studies?"

Translation: AAV9 is the virus that delivers the gene therapy. GFP stands for "green fluorescent protein," which comes from jellyfish. When the GFP gene is attached to any other gene, the corresponding protein lights up green—which a researcher can actually see when the tissue where the gene is used is viewed under a microscope. The pig experiment used GFP, rather than gigaxonin, to see if the virus could safely shuttle a gene into the spinal cord, but the fluid delivered with the virus had killed some of the pigs. It's better to work out the delivery details first on pigs and monkeys before attempting the procedure on a child.

Lori continued before Chris and Karen could comment. "SMA is so common, second next to cystic fibrosis. We're ultrarare. SMA has millions of dollars. If we aligned . . ." she pleaded, near tears.

Karen was quiet a few moments, her chin on her hands, thinking. Then she straightened up and smiled. "Our scientific advisory board meets at the end of July. We haven't decided yet what to do about gene therapy. Like you, we don't really want to be the first." She paused as Lori and Steve leaned closer. "But our meeting is July twenty-third to twenty-fifth. You should come." Lori let out her breath and broke out into a huge smile, then hugged Steve.

Because GAN is so rare, Lori and Matt face a more difficult task than does the SMA community. From Tricia and Phil Milto they learned how to prepare for the uphill battle of attracting interest in an extremely rare disease. Two of the three Milto boys have Batten disease, aka juvenile neuronal ceroid lipofuscinosis. It, too, is autosomal recessive. It, too, is extremely rare: only 1 in 200,000 newborns will develop it.

The "first sign of the nightmare," recalls Tricia, was when four-year-old Nathan suddenly couldn't see in a darkened movie theater, and soon after began tripping and colliding with things. Doctors ruled out several conditions, including Corey's LCA2, and after several false leads, diagnosed Batten. "The doctor said we should just go home and enjoy the rest of the time we had left," Phil says, still bristling at the advice to give up and give in. Then the Miltos' youngest son, PJ, developed symptoms.

Phil tapped his entrepreneurial skills and became a fixture at scientific meetings, while Tricia used her PR and fund-raising experience to organize. Years before Lori met with the SMA researchers to talk about their mouse cure, Phil Milto sat at a similar hotel bar at a scientific meeting with a researcher who had cured Batten's in mice, asking if he would help with a gene therapy for children. He did. The Miltos' organization, Nathan's Battle, raised more than $6 million for the first clinical trial for Batten disease, conducted at New York-Presbyterian/Cornell Medical Center in New York City. Both Nathan and PJ took part. Although the gene therapy did seem to slow disease progression, the project could no longer keep raising the millions of dollars to keep going.

Now, years later, a new and improved gene delivery system—the one Corey had—is being tested in children with Batten disease, again at Cornell Medical Center. One of the beauties of gene therapy is that the exact function of a gene need not be fully understood for its delivery to help. In Batten disease, children are missing a protein called CLN3, which stands for "ceroid-lipofuscinosis, neuronal 3." How it causes the disease when missing or abnormal remains a mystery. The protein is found in many parts of many cell types, but researchers think that in brain neurons, its absence or inactivity causes molecular debris to accumulate in the lysosomes, which are tiny sacs that function in cells as garbage dumps. Researchers send normal versions of the gene cradled in 900 billion or so viral vectors into the children's brains through six holes drilled into their skulls. So far it works, says Ron Crystal, director of the Belfer Gene Therapy Core Facility at Cornell Medical Center. Cells secrete new proteins, which can help neighboring cells, too. The protein need restore only 10 percent of the cells to see an effect, Crystal says.

The new Batten disease gene therapy trial is too late for the Milto brothers. But the positive results with the new viral vectors may be good news for Hannah and the other kids with GAN.

~

In their frantic first days of googling, Lori and Matt discovered websites dedicated to children with all sorts of terrible diseases, and also the umbrella organizations—the National Organization for Rare Disorders, Genetic Alliance, the Children's Rare Disease Network. They also—as do most parents of children newly diagnosed with an extremely rare disorder—reached out to find other families living the same nightmare. Lori and Matt quickly located GAN patients in Germany, India, Canada, New Zealand, and Australia, as well as the United States. And they actually found some encouraging news: patients *can* live into their twenties, with normal intellect. (They discovered that in the few families in which the children were intellectually disabled, it was due to other recessive mutations passed on from parents who are blood relatives.)

Lori and Matt struck up friendships with the other GAN families, but it was difficult to glimpse what lay ahead for Hannah. "A woman in Florida cares full-time for two sons in their twenties. They are quadriplegic, with feeding tubes and ventilators, but they can move and blink. They respond. They are cognitively intact," says Lori, who talks to the mother often. When Lori watches Hannah push her cousin's dog around in a stroller at the race, looking like a normal little girl, she finds it hard to imagine a future of immobility and total dependence for her.

The Sameses organized the global effort in GAN gene therapy, continually contacting researchers and searching for grant opportunities, as affected families responded to the Hannah's Hope website and provided both old-fashioned genealogical information as well as results of DNA tests. Problems arose. Divorced parents were unwilling to cooperate enough to get DNA results to Lori. People called with incorrect self-diagnoses. Once Lori managed to determine which respondents actually had GAN, she compiled lists of documented cases that included as much information as possible, down to the specific DNA bases that were affected in particular families. Meanwhile, the number of identified GAN mutations hit twenty, causing different severities of the disease. Lori and Matt discovered that they each have a gigaxonin gene missing the exact same stretch of DNA. Considering how rare the disease is, this is a huge coincidence. One explanation is that Lori and Matt are distant relatives and inherited the same mutation from a shared ancestor, such as a great-great-grandparent that they unknowingly have in common. Hannah, therefore, has the same mutation in both copies of the *GAN* gene. Corey's situation is more common—a different mutation from each parent. (Although gene therapy can work on any mutation, certain types of mutations may be treated with drugs that make the cell ignore the glitch and produce the protein anyway. That is one reason why Lori and Matt, and the other parents, were tested.)

While the tiny GAN community came together and shared information and DNA, Lori mastered the trajectory for drug approval, for gene therapies are considered drugs.

First come the "preclinicals"—experiments with cells growing in dishes and with animals that either naturally have a disease that people also have or are genetically modified to harbor a mutant human gene; for example, Yanmin Yang's GAN mice are stand-ins for children. About 80 to 90 percent of drug candidates fail at the preclinical stage, usually due to toxicity—the drugs don't work at levels that don't kill or injure the experimental subjects. This high percentage of failure is why drug developers call the preclinical stage the "Valley of Death." It can take $10 million and two to four years to crawl up and out. When giving presentations on drug development, researchers typically show a slide depicting two cliffs on either side of a chasm, with one mountain labeled "basic research" or "NIH" and the other labeled "Drugs" or "FDA." "Preclinical studies, aka the Valley of Death, are where projects go to die," says Francis Collins, director of the NIH.

Lori, however, plans to leap from one mountaintop to the next with clever scheduling. She compared various deadlines for grant and regulatory submissions and determined that the GAN team could be ready to submit their "investigational new drug" application—the IND, which comes at the end of preclinical trials, when the investigation is ready to turn to people—in eighteen months. The accelerated timeline that she was going for, thinking ahead to avoid downtime between preclinical and clinical trials, echoes a changing mind-set at the two agencies, as well as among researchers, who are trying to take several steps in parallel rather than in traditional series. "The long timeline to obtain multiple awards for funding successive stages of research, and the redundant regulatory reviews, present nearly insurmountable hurdles to progress," says Don Kohn, who is conducting the SCID-X1 trial in Los Angeles.

As preclinical tests are under way or winding down, the "pre-IND meeting" with the FDA takes place, allowing troubleshooting for the planned clinical trial. Hannah's Hope Fund had several pre-pre-IND meetings. Once the IND is submitted, a thirty-day review ensues, and if the agency gives the go-ahead, phase 1 of the clinical trial begins. In it, a few individuals take escalating doses to assess safety, side effects, and the way the drug acts in the body. Phase 1 is the part that the media so often misunderstands in "therapeutic misconception"—it isn't meant to cure, or even treat, but instead to determine safety and biological activity. A phase 2 clinical trial expands the number of participants and looks for efficacy; phase 3 builds the numbers. Like a contestant in a reality show, a drug candidate becomes more likely to win as it advances along the pipeline. "The IND is the entry point to the clinical trial. After phases one, two, and three, the FDA says, 'Yes, it's okay to market that,'" explains Collins.

Getting through phases 1, 2, and 3 is hard for any drug, but daunting for one intended to treat an exceedingly rare condition. The statistics on very rare diseases reflect the extra obstacles they have traditionally faced in the drug development process—in addition to the mysteries of science, it is very, very expensive. A rare disease, according to the FDA, is one that affects fewer than 200,000 individuals in the United States at any given time. And many rare diseases are far rarer than that. But taken together, the rare diseases number more than 6,800, and they affect many millions of people. Only about 200 of them have treatments, and the reason is economic: it costs hundreds of millions of dollars and typically takes eight to twelve years to bring a drug to market. How can this level of development be supported by an end product that treats only a few patients? And gene therapy presents an additional hurdle—it is, ideally, a onetime treatment, not a continual source of income like a statin or antidepressant.

Because of the unattractiveness to biotech and big pharma of treating rare diseases, the FDA and NIH are making special efforts to help the rare disease community, especially the unicorns among the zebras. For example, both agencies are lowering hurdles to repurpose drugs, an increasing practice as genome research is revealing unexpected links between diseases. Rather than competing, rare disease researchers are now sharing their discoveries, so a patient doesn't waste time trying a drug that others have already found to be toxic. In addition, clinical trials with very small numbers of participants commonly combine phases 1 and 2 (as did Corey's trial), testing safety while hoping for efficacy, shortening the time to build the numbers that the FDA wants to see to approve a treatment. The NIH-FDA Joint Leadership Council helps with special situations that may arise from trying to develop treatments for rare diseases. "This program gives FDA the science background to assess unusual clinical trial designs. For example, a disease may be so rare that there are too few people to do a phase three trial," says Collins. The FDA has just kicked off a five-year plan to expedite approval of treatments for rare diseases. An initial public meeting was held in October 2011 to gauge interest, and the Center for Drug Evaluation and Research's Rare Disease Program, which will provide specialized training on drug development for rare diseases to scientists and clinicians and will extend outreach to patient-run organizations, is slated to follow in late 2012.

The FDA five-year plan is in response to the pharmaceutical industry's finally beginning to pay attention to the rare diseases. Part of the impetus is that many of the blockbuster drugs spawned during the golden age of pharmaceutical innovation of the 1990s are going off-patent—six of the ten top sellers by 2013, including the megamillion generators Lipitor and Plavix. Giants such as Pfizer, GlaxoSmithKline, Novartis, and Eli Lilly are starting rare disease treatment initiatives. At the same time, new organizations and programs are moving the mountains that surround the Valley of Death by shifting phase 1/2 trials away from big pharma and to academia and the small biotechs, so by the time phase 3 rolls around, risk is lowered and the big guns are more likely to take on a project. In the United States, the new National Center for Advancing Translational Sciences, and in the UK, the Medical Research Council's Developmental Pathway Funding Scheme, and Wellcome Trust's Seeding Drug Discovery Initiative, are accelerating the clinical trial process, and the NIH and the European Commission have formed the International Rare Disease Research Consortium. Its ambitious goal: a diagnostic test for each of the 6,000-plus conditions and treatment for 200 of them by 2020. And U.S. Senator Bob Casey introduced the Creating Hope Act of 2011, to seek new treatments for rare and neglected diseases that disproportionately affect children.

Lori put together the GAN scientific dream team with the FDA road map and these new efforts in mind. Unimpressed by titles and power, she sensed that fresh ideas and key contributions can come from a graduate student, postdoc, or young investigator, someone whose enthusiasm is high and sleep requirements low. After intense months of networking, Hannah's Hope Fund held its first scientific meeting in Boston in August 2009. Recalls Steve Gray, "I was impressed with the panel of twenty scientists Lori assembled. Some were fans of gene therapy and some not, but all agreed, if there was any hope of a solution in a reasonable time frame, gene therapy was going to be it."

Lori knew the science and trusted her instincts. Gray, the young molecular biologist, would be the one to engineer the viral vector that would go into Hannah and several other youngsters. When Hannah's Hope Fund offered to support his research on GAN, Gray was stunned. "I was very much a junior investigator. To take on a project of this magnitude . . . at a typical granting agency I'd never be a lead investigator. Lori caught me when I was wide open to ideas." And he knew just what he'd do first: get the gigaxonin gene into a safe AAV vector and give it to mice missing the gene. "We were starting from scratch. Hannah's Hope was taking a big risk."

Today Lori and Steve are fast friends. They both smile, remembering their first phone conversation after the meeting in Boston.

"I was at my parents' house, and we talked for an hour. When he said he had a daughter Hannah's age, I knew it was meant to be. He had the scientific background *and* the passion, the personal connection," says Lori.

Steve responds in a low voice. "When I learned about GAN, I put myself in Lori and Matt's shoes. And I would go for it, gene therapy, if there were no other options. If there was enough science and low enough risk/benefit, I'd do it."

Hannah's Hope held more meetings and fund-raisers, as e-mails flew back and forth among researchers, with Lori the nexus. The team began to jell: the vector builder, the designer of the gene cargo and its controls, the stem cells that would fill lab dishes with hard-to-grow human motor neurons, the makers of mutant mice, and, finally, the neurosurgeon to deliver the therapy to pigs and then to people. Meanwhile, Hannah and a few other kids had baseline tests of their speech, cognition, and fine and gross motor skills, as well as spinal taps, brain scans, and nerve conduction tests.

The GAN gene therapy plan also needed a "readout," a way to assess efficacy (a response as spectacular as Corey's at the zoo is most unusual). In general, trial design must follow an observable or measureable change that has a clinical effect. Does the intervention slow or stop the disease, or does it only mask symptoms as the pathology plods on? Lorenzo Odone's blood levels of very long chain fatty acids indeed plunged when he took the eponymous oil, but did he feel better? He did in the Hollywood version, when he dramatically moved his pinkie as the camera zoomed in for a close-up. The real Lorenzo lived far beyond expectations, but even his father doesn't know if the oil was responsible. Will Hannah post-gene-therapy not only make gigaxonin but be able to walk?

Lori found that scientific research and the regulatory road map don't address one key problem in working toward a clinical trial—reputation and hence funding. The most difficult organisms to work with, she discovered, were competing researchers. Some scientists hesitate to talk about experimental results until the data are safely published, their turf marked in the pages of *Science* or *Nature* like a dog peeing on a log. Some quibble about the order of names on a paper. Lori was terrified when she heard that one key team member had delayed disclosing findings that would have saved the other researchers weeks of work because he didn't like the position of his name in the author list. She feared that the loss of a few weeks could translate into a year's setback as grant deadlines passed, and one was looming—the NIH's Rapid Access to Interventional Development program, RAID (recently renamed BrIDGs), which would underwrite pig and monkey toxicity studies. Without RAID funding, Lori worried, Hannah's Hope Fund might run out of money just as the clinical trial was getting off the ground. Plus, RAID could find a company or foundation to pick up the costs of the clinical trial. (RAID supplements the Orphan Drug Act of 1983, which incentivizes industry to develop drugs for rare disorders. The term *orphan* has been replaced with *rare*.)

When a parent is planning a clinical trial for the ticking time bomb of a genetic disease, delays to foster a scientific career are simply intolerable. Lori, because her organization was supporting the research, tapped out blistering messages on her BlackBerry to the European researchers and raged over the phone to the U.S.-based ones, and eventually the lines of communication opened. All of the work on mice and pigs, and the legacy of viral vectors in other gene therapy trials leading to AAV9, have put the pieces in place for Hannah's gene therapy. It will be soon.

Lori Sames is at her most amazing at a scientific conference.

A scientific conference is out of the realm of experience of most ordinary people. The intensity and sense of urgency is stronger than an annual meeting of, say, real estate agents or bathroom fixture salespeople. At the annual American Society of Gene and Cell Therapy meeting in 2010, that sense was pervasive. Finally, twenty years after the first gene therapy on four-year-old Ashi for ADA deficiency had led to confusion over whether the approach had actually worked or the supplemental enzyme replacement had, the technology's time had truly come.

"You can feel the excitement here, with changes at FDA, enough safe viruses and no fear of cancer," murmured Steve Gray as he looked around at the knots of researchers updating one another at the exhibition. The meeting, at a sprawling hotel a short walk from the National Zoo, started with a two-day workshop on navigating the FDA obstacle course. Then came four days of reports on research results, most too new to have hit the journals.

Lori sat through it all. But she didn't do much actual sitting.

There's so much science to pack into the few days of a conference that concurrent sessions are a must. And so scientists zigzagged through the hotel's hallways, pinging from meeting room to meeting room like mice in a maze, trying to be at more than one talk at a time, to take it all in. In the larger conference rooms, huge paired screens displayed the insides and outsides of the various viruses used to shuttle genes into cells. Diagrams of multicolored rectangles peppered with abbreviations of gene names—*tat, env, E4*—riveted the attendees, appearing as gibberish to the staff replenishing the coffee. In some lecture rooms, split screens showed unfortunate rodents in various stages of decrepitude on the left, recovering lost looks and athletic abilities after gene therapy on the right. Other slides showed their brains, sliced up like salami, with colored dots marking where genes had hit their targets.

When Lori wasn't in the front row of a session, or chatting up the speaker immediately after, she was at the bar, or at a restaurant, with the top scientists, networking nonstop. She corraled one of the founders of the field and extracted a promise to meet in the lobby when the sessions ended for the day.

Promptly at 5:00 p.m., Lori sat down with the researcher. She was friendly and open, explaining all that Hannah's Hope Fund had accomplished. She talked about AAV and knockout mice and catheterized farm pigs, easily slipping into the lingo of "routes of administration" and "biodistribution studies." She told him about the team members, from institutions in North Carolina, Atlanta, New York City, and Paris. He listened and offered to help, but she felt he was at the same time dismissive and condescending. "You're underestimating the complexity and difficulty. Delivery into the cerebrospinal fluid will only get the superficial cell layers. Does anyone you are working with have any clinical experience? Are you dealing with anybody who has done this before, taken a drug all the way through the FDA?" Multi-institution teams don't work, he added, and she must be wary of biotech companies. "I know, I founded one," he said. Preparing the FDA documents, he explained, requires specific expertise.

Lori listened patiently. Nothing can rattle her, not even a famous scientist explaining the obvious as if talking to a preschooler. Then the researcher dramatically took out a thick document and slowly pushed it toward her on the small bar table. "See, we're already working on GAN." Lori glanced at it. But instead of peppering him with questions, as he clearly expected, she ignored the report and returned to her story.

She'd already seen the paper, a proposal for a GAN clinical trial, sixteen months earlier. Because nothing had obviously gotten under way since then, it was clear that he was bullshitting her, marking his turf, unaware that she was onto him. The conversation that had gone nowhere wound down, and Lori thanked him and moved on.

Lori took a few minutes off to bolt down dinner in the hotel restaurant, which overlooked the escalator leading down to the exhibition. Mid-bite, she spotted the researcher who sequenced her and Matt's DNA, standing at the top of the escalator. Pushing aside her plate of half-eaten chicken, she was off, catching up to him before he even got off the escalator. She took his arm and guided him into the large room where lab equipment vendors had set up tables alongside representatives from biotech companies and a few government agencies. Conversation with the DNA-sequencer wrapped up, Lori approached a friendly-looking physician representing the National Heart, Lung, and Blood Institute's Gene Therapy Resource program. Once she had the woman's attention, Lori launched into her most immediate concern, always the most immediate concern of a parent of a child with a rare disease—*time*.

Her plight poured out in a staccato of pleas: "Our phase one trial will have nine kids because GAN's so rare. We have to pay for toxicity testing and development of the vector. We can't afford a delay of one year. By the time we have the data, the funding cycle would be so far out that the entire study would take seven years. Kids don't have time for the NIH RAID program to pay for toxicity testing. We don't fit the NIH schedule. These kids need to be treated. They are going to die."

The NHLBI doc was sympathetic. "Yes, the time issue always comes up. An ancillary trial is faster, but it sounds like your disease is too rare," she said sincerely. "But you can submit data as you go." This turned out to be a valuable contact, for come September, when Hannah's Hope Fund missed the RAID deadline, Lori knew to ask for an extension—and got it.

The high point of the gene therapy meeting, for Lori and the other thousand-plus attendees, came on Friday night, at the presidential symposium. A nine-year-old boy walked onstage holding the hands of his nervous-looking parents. A woman with flowing reddish hair held in place with pink barrettes and wearing a short jumper with red tights followed—Dr. Jean Bennett. As the crowd sat stunned, the family settled into their seats, and with a broad grin, Dr. Jean said, "I'd like to introduce the youngest person ever to speak at ASGCT—this is Corey Haas."

With that simple sentence, she brought many in the crowd of usually staid scientists, especially the older ones, to tears. Seeing Corey, hearing Corey, laid to rest the doubts over Ashi and Cynthia, the anguish over Jesse and Jolee and the leukemia boys. Gene therapy was and is, finally, a resounding success.

Dr. Jean then told the story of Corey's gene therapy, including the now-famous film of the boy navigating the mobility course with his treated eye (in seconds) and then his untreated eye (in several looooong minutes). Then Corey calmly answered questions from the audience. Ethan beamed; Nancy dabbed at her eyes.

"Corey, would you like to have your second eye done?" Dr. Jean asked.

"Sure!"

"When?"

"How about tonight?"

The crowd roared. Corey, a veteran of news and talk shows, was already a pro at handling an audience.

After the Q and A, when the family returned to their seats in the front row, Corey was mobbed. A woman scientist bent down and threw her arms around him, crushing him to her pillowy bosom. "You're going to be a scientist when you grow up, aren't you?" she asked as Corey tried politely to squirm away. He turned around and broke free for a moment, only to be stopped by a distinguished older man thrusting out his hand. "You are the bravest person I know," he said, shaking hands, and Corey grinned.

On the fringe of the crowd stood the usually laughing Lori, uncharacteristically still and silent, tears streaming down her face. For if gene therapy can cure Corey, it can cure Hannah, too.

The next afternoon, when Lori stepped onto the sidewalk outside baggage claim at Albany International Airport and scooped up Hannah as she stumbled toward her from the family van, the little girl was unimpressed.

"Mommy, why do you have to go away to learn science?"

PART V

AFTER GENE THERAPY

I know in my heart that the gene therapy has kept Jacob from slipping into a vegetative state. He is with us, he engages us, he laughs at what is funny and responds appropriately when we talk to him.

—Jordana Holovath

12

AMAZING WOMEN

By all accounts, Lindsay Karlin was an exquisite baby. She was so pretty that when she lay in her stroller, strangers would admire her delicate features and her porcelain beauty. "Oh, she's sleeping," they'd coo. But one day in response to yet another well-wisher, the little girl's mother, Helene, smiled back and calmly said, "She's not sleeping, she's dying."

But Lindsay Karlin didn't die, likely thanks to gene therapy.

The memory of her mother's disturbing remark to the well-meaning stranger is still so strong to Molly Karlin, who was twelve years old at the time, that she recalls that it happened in an elevator. She was embarrassed then, but today she understands. "It was hard for her to have people look at Lindsay and think she was normal."

For Lindsay's first few weeks of life, all had seemed fine. "She's one of the healthiest babies I've seen in a long time!" exclaimed the obstetrician at Lindsay's birth on July 18, 1994. But by the baby's seventh week, Helene, a PhD in psychology who already had two children, feared that something was wrong. Lindsay's eyes weren't tracking. When Molly held out a stuffed toy to her baby sister, Lindsay's eyes would flick back and forth, but never focused on the brightly colored animal. The girls' father, Roger, an internist, suspected that his beautiful baby couldn't see.

By three months old, when Lindsay wasn't smiling as her sisters, Samantha and Molly, had, Helene and Roger grew more concerned.

Around Halloween, Lindsay saw an ophthalmologist, who prescribed glasses so thick that she looked like a court jester to Molly. The eye doctor, concentrating on one body part, thought the problem was only visual. But the worried parents saw more, and asked for a referral to a pediatric neurologist.

Tests followed, and the phone call came about a month later. Molly says it was the worst moment of her life.

"It was a Monday night, about eight o'clock, and it was raining. I picked up the phone, and someone asked for my dad. He wasn't home, so I gave it to my mom. She listened for a minute and then started crying. I started to cry, too, and gave her a paper and pen and she started writing as she listened. Someone from a genetics lab had called back with urine test results and said it was Canavan disease. 'Is Lindsay gonna die?' I asked my mom. And she said yes. The next day I went to the school library and looked it up. The medical encyclopedia had four lines on it, and called it 'spongy degeneration of the brain,' and used the word *demyelination*."

A few days later, the family went to Yale University, about an hour's drive from their home in New Fairfield, Connecticut, to meet with specialists. "The neurologist said, 'Your child has a terrible degenerative disease and she is not going to live past three, so think about placing her at a facility. And get her and her sisters tested for the mutation,' recalls Helene. Even at her young age, Molly recognized the doctor's insensitivity. "My parents weren't going to take this lying down."

In the aftermath of the diagnosis, Roger followed in the footsteps of the Odones and the Salzman sisters by plunging into research. Although Roger had practiced medicine for fifteen years, he'd never heard of Canavan disease, which is known to affect only about two hundred people in the United States. He quickly learned that in the cells of people with Canavan's, there's not enough of an enzyme called aspartoacylase (ASPA). Certain brain cells need ASPA to break down a biochemical called NAA (N-acetyl aspartic acid) into pieces used to make myelin, the fatty sheath of the nervous system. The excess NAA spills into the urine, providing a handy biomarker to help diagnose the disease. This biochemical mayhem was turning the white matter of Lindsay's brain into a spongy mass of fluid-filled bubbles. It's the same end result as in the "Lorenzo's oil" disease, ALD, but with a different cause and a much earlier onset. ALD at least allowed kids such as Lorenzo Odone and Oliver Lapin a few years of normal childhood.

Children with Canavan disease rarely surpass the skill levels of infancy, remaining trapped in utterly helpless, yet growing and developing bodies. Most never speak, walk, or even turn over during sleep. But parents of Canavan kids have an uncanny feeling, an inner conviction, that someone is in there. They can see it and hear it in smiles and laughs and the occasional *ahhh*, all eerily fitting the situation. It's as if the relentless pathology spares the very parts of being human that attract and hold love. A little boy giggles when his father makes a fartlike noise; a little girl stops crying and beams when a friend plays guitar and sings, moving her fingers in front of her as if playing a keyboard. Lindsay Karlin clearly adores her dog, a Maltese named Snowflake. "I take her hand and ask, 'Do you want to pet the dog?' and she goes *oooooh*, and I help her," says Roger. Many Canavan kids understand spoken language, and struggle to indicate that they can do so. For example, Lindsay's parents noticed that as a preschooler she opened her mouth to be fed when they asked if she wanted to eat, in English, and also when the housekeeper asked in Spanish. The little girl learned.

As was the case with Corey's and Hannah's conditions, a diagnosis of Canavan disease can be missed, because it is so out of the experience of most doctors. Canavan disease is more common among Ashkenazi Jewish people, among whom 1 in 40 is a carrier. (Helene and Roger Karlin are both carriers.) Diagnosis may be swift if a doctor recognizes the warning signs of a big and lolling head and floppy muscles. Lindsay was as flaccid as a rag doll, Helene recalls.

But several disorders have these signs, and even the sponginess that appears on a brain scan or in a biopsy is not unique to Canavan disease. High NAA in the urine strongly suggests Canavan. Lack of ASPA clinches a Canavan diagnosis, but that requires a DNA test because the enzyme isn't normally in the blood.

Roger's connections in the medical community were useful in helping him discover what was wrong with little Lindsay. And then geneticist after geneticist advised him to give up and prepare for the impending loss of his beguiling daughter. It would be easier on the family to institutionalize her, they kindly said. To his horror, Roger learned that some parents let their children die from a seizure at home, rather than take them to the ER, because they are distraught from a recent diagnosis, exhausted from living for years with the disease, or want to end their child's suffering. "My dad isn't one to quit, and he instantly started trying to find a place with treatment. He called hospital administrators, doctors, geneticists, everyone under the sun," Molly recalls.

After about six weeks, Roger made a fateful connection. "I called the NIH and spoke to a gentleman named Roscoe Brady, who had come up with an enzyme replacement therapy for another single-gene disease, Gaucher's disease. He knew a scientist, who he called a maverick, working on gene therapy—Matt During." Conveniently, During, an MD and PhD, was director of the Laboratory of Molecular Pharmacology and Neurogenetics at Yale. The geographical coincidences continued when a family living nearby contacted the Karlins after reading a newspaper article about Lindsay. The Mushins had a little girl just a few months older than Lindsay who also had Canavan disease. Their daughter, Alyssa, and Lindsay would be the first children to undergo gene therapy for it.

Roger and Helene read a report that Matt During and his associate, the postdoctoral researcher Paola Leone, had just published in the prestigious journal *Nature Genetics* detailing gene therapy for rats with Parkinson's disease. The researchers were young, thirty-seven and thirty-two, respectively, but already very experienced, knowledgeable, and excited about translating scientific discoveries into clinical practice. From many phone conversations, the Karlins felt that Drs. During and Leone, who quickly became Matt and Paola, were the right team to tackle Canavan. So in May 1995, the family visited the During lab at Yale for the first time. And it was love at first sight. "Roger went to talk to Matt and I talked to Paola at length. I won her heart. She felt so sorry for us. They both said working on Canavan gene therapy would be extremely challenging and yet very rewarding. They'd be working not with yeast, not with mice or monkeys, but with real people. Soon they had Lindsay's and Alyssa's pictures up all over the lab," Helene recalls.

The two researchers were intrigued. Canavan disease seemed a good candidate for gene therapy, and perhaps could pave the way for treatments for common neurodegenerative diseases, such as Parkinson's and Alzheimer's. The major stumbling block was the same: getting the gene where it was needed in the brain, and keeping it away from other parts. Treating Parkinson's disease with L-dopa, for example, leads to another condition, tardive dyskinesia, because the L-dopa also goes to brain areas that cause uncontrollable movements different from those of the disease it is meant to treat. Paola had several reasons to take on Canavan: "It's caused by a single gene, and a single organ is affected, although unfortunately it is the most complex organ, the brain. And there's no treatment whatsoever." Also, the fact that carriers such as Helene and Roger, with half the normal amount of the enzyme ASPA, nevertheless enjoyed good health, meant that just partly restoring enzyme function might help the sick children.

Other researchers in the world are working on Canavan disease, but over the years, it has become more or less Paola Leone's baby. The photos and drawings of her young patients in her lab attest to the fact that they are her life. She has no children, but many pets fill her heart with happiness—the usual dogs and cats but also a talking and singing parrot, a horse, and a donkey. If she had a child, she says, she'd be so mesmerized she'd just sit and stare at him and not get any work done. The Karlins' plight got under her skin. "Little Lindsay deeply affected me. I said to myself, 'Don't do it, the disease is very rare. There's no funding, there's not even an animal model to test a gene therapy.' On the other hand, what other chances do children like Lindsay and Alyssa have? So I chose to take the difficult path to move science forward in time to make a difference." She has dedicated her life to it.

The human brain, especially the cortex, the seat of individuality, intelligence, and personality, has been Paola Leone's passion since her childhood in Sardinia. "I was more interested in science than medicine. If I were a physician, I knew that I couldn't study the complexity of the brain. When I earned my neuroscience doctorate in 1987 from the University of Padua in Italy, we didn't know much at all about the cerebral cortex. We still don't." Paola also passed on medical school because she couldn't deal with death and dying, and veterinary school was out because she couldn't handle large animal dissection. So although she wasn't trained to work with patients, her lack of a medical degree hasn't kept her from tackling Canavan. She has the smarts to figure out how a disease devastates the human brain, the compassion and love to be a source of constant support to the families, and the training and experience to work with highly specialized physicians.

Even fifteen years after she met Lindsay Karlin, Paola remains overwhelmed by the experience. "Destiny chose me. I never imagined I'd do this. I have no religious ties, but I'm very spiritual. I feel I was guided to meet a family with a child with Canavan and begin this journey."

Paola Leone is not the only visionary woman who is part of the Canavan story. The disease takes its name from Myrtelle Canavan, one of the "forgotten women" of biomedical science.

Born in 1879 in Michigan, Myrtelle Canavan received her MD in 1905 from the Woman's Medical College of Pennsylvania. She then held a series of positions in pathology at state hospitals in Massachusetts. At that time, pathologists in the United States numbered only about fifty, and they tended to be specialists in anatomical oddities, rather than their modern counterparts, who are expert in histology, cell biology, and biochemistry. In 1924, Dr. Canavan took a post at the Warren Anatomical Museum at Harvard Medical School while also teaching at Boston University and the University of Vermont College of Medicine. At first she was assistant curator, because the museum committee doubted that a woman could handle the full title, although she did the full job. Eventually, she became curator.

Myrtelle Canavan liked to sit and stare at dissected brains and splayed spinal cords, trying to link peculiarities that she could see in the body parts with what they did to the people who donated them. At that time, patients with nervous system disorders were more likely to receive a vague diagnosis of "feeble mindedness" or "imbecility" than a specific disease or syndrome. Dr. Canavan sought better explanations. She also studied the effects of infections on the brain, as well as the mysterious white deposits in the brains of people who had had multiple sclerosis. In 1931 she discovered the disease that would bear her name, writing in the *Archives of Neurology and Psychiatry* about the oddly "spongy" brain of a child who had died at sixteen months, still as helpless as a newborn.

Harvard Medical School never granted Myrtelle Canavan full faculty status. She acquired more than fifteen hundred specimens for the museum, continued teaching and mentoring, trained the man who went on to train most of the nation's neurosurgeons, and published prodigiously. She retired from the museum in 1945, and died in 1953 of a brain condition with a name, Parkinson's disease.

In 1949, a paper in a medical journal from Belgium reported the same condition as the child in Canavan's case study in three babies, noting something else the children had in common: they were all Ashkenazi Jews, from families that came from eastern Europe.

Then another study found that of forty-eight newly diagnosed patients, twenty-eight were not only from Jewish families but came from a very specific area—Vilnius, Lithuania. In that city existed, for two miserable years at the start of the Second World War, a ghetto that was once home to about sixty thousand Jews.

13

THE JEWISH GENETIC DISEASES

THE ESTABLISHMENT OF THE VILNA GHETTO WAS BOTH planned and methodical. On August 31, 1941, the Nazis staged a sniping of two German soldiers from the window of a Jewish family's apartment in the poor part of the city. Soldiers dragged two Jewish men out of the apartment onto the street and shot them in front of a gathering crowd. The soldiers then went on a rampage, driving people out of their apartment buildings, killing many as they ran. The next day, German soldiers forced Jewish women and children all over the city from their homes and stuffed them into the newly created ghetto in the vacated poor section. They rounded up the Jewish men at work and sent them to the ghetto too. Then, from September 1 to September 3, a mere two days, the Nazis assembled ten thousand of these people and sent them to a prison, where they were murdered. The surviving Jews in the city, as well as non-Jews married to them, were herded into the ghetto.

The Nazis split the Vilna Ghetto into two parts so that they could easily watch the goings-on from a strip down the middle. By November, the smaller of the two ghettos shrank and then disappeared as its residents were systematically killed. In the larger ghetto, only slave labor was permitted to survive, as the Nazis killed the old and young, the sick and starving. By September 1943, most of the survivors of the Vilna Ghetto were sent to other camps, leaving only about 250 prisoners behind. Some of them escaped to the surrounding forests and survived, and some of those eventually made it to the United States. That small group of immigrants, exact size unknown, brought with them the two mutations that cause 97 percent of the cases of Canavan disease in the Jewish people of Ashkenazi background today.

Geneticists have a name for the tendency of an otherwise rare single-gene disease to be more common among certain population groups, due to a mutation brought in with original settlers: a founder effect. It is, in essence, a sampling error. Had the survivors who fled the Vilna Ghetto *not* included individuals with a Canavan mutation, the condition would not be one of the dozen and a half "Jewish genetic diseases" recognized today. People in a population marrying only among themselves amplify a founder effect, as shared disease-causing gene variants are passed from generation to generation. When two copies show up in a person, the recessive disease results. The mutations have nothing to do with religion and everything to do with human behavior. They persist thanks to genetic isolation due to choosing marriage partners from within a group, a behavior catalyzed by persecution.

The Jewish people have a long history of being ousted from their homes, if they were lucky, or murdered there if they were not. Jews were first kicked out of Israel in 723 B.C.E., settling mostly in what is now part of Turkey. They were forced out again in 586 B.C.E., and settled in Egypt and Iraq, then called Babylonia. By the time of the Roman Empire, the Jewish people numbered about 8 million. In the first century A.D., they moved westward into Europe after the Romans evicted them. Then, in the eleventh century, the Crusades slashed the population, forcing the survivors into eastern Europe, where they joined other Jews who had left Turkey. Today, 80 percent of Jews in the United States are Ashkenazi. Their ancestors came from western, central, and eastern Europe, especially Poland, Romania, Russia, and Germany. Most of the rest of U.S. Jews are Sephardic, coming from the Mediterranean, primarily Spain, Libya, and Morocco. Others are from Greece, Iraq, and

Iran. Analysis of the genomes of many modern Jewish people reveals that almost any two are, unknowingly, fourth or fifth cousins. One in four Ashkenazi Jews is a carrier of a Jewish genetic disease. The combination of Jewish people marrying among themselves and being herded into ghettos as their numbers plummeted maintained recessive mutations that were passed along, silently, in healthy carriers.

On a large scale, the Jewish population has repeatedly crashed and re-formed. Each time it sprang anew from a small group of survivors so that, eventually, the same recessive mutations began showing up in those descendants. This ebb and flow of the Jewish population, sequestering and then amplifying specific mutations, is responsible for the higher numbers of Jewish people today who have Canavan disease, Bloom syndrome, cystic fibrosis, Gaucher disease, familial dysautonomia, maple syrup urine disease, or any of a dozen others. Although these conditions are also seen in non-Jews, the specific *mutations*, the precise changes in the gene, that are common among Jews are rare in other groups. Tay-Sachs disease, once the quintessential Jewish genetic disease, is no longer on the list because it's almost gone, at least among Jewish people.

Modern Israel serves as a living laboratory to study the genetic consequences of the serial strangulation of the Jewish gene pool, because the nation is home to varied groups of survivors; their communities hold a trove of useful genetic information. Researchers comb disease registries for otherwise rare disorders that are common in certain Israeli communities, then compare statistics from the founding population to the much smaller contemporary Israeli one. For example, researchers from Hadassah-Hebrew University Medical Center and colleagues discovered that a distinct mutation causing Leber congenital amaurosis type 2 (LCA2, Corey's disease) was quite common among Israeli families from North Africa, particularly Morocco. They calculated that the mutation found in today's Israelis began about 153 generations—or about

3,825 years—ago. This type of information can help physicians diagnose diseases more readily, by identifying rare conditions that are more common in some populations.

The Jewish genetic diseases are, happily, on a sharp decline. Some of this is due to the dilution of ancestry that comes when Jews marry non-Jews. But the vanishing diseases also reflect a concerted effort, and one that doesn't involve sophisticated technology. It sprang from a rabbi in Brooklyn who had a great idea, spawned from a personal tragedy. With the sequencing of the human genome and resulting great expansion of carrier tests, the rabbi's approach is today preventing many genetic diseases.

In 1965, when the Holocaust survivor Rabbi Josef Ekstein and his wife had their first child with Tay-Sachs disease, hardly anyone had ever heard of it. The boy seemed well for six months, but then he lost muscle tone and his small body became racked with seizures. Infection followed infection, one blending into the next as his immune system struggled to fight back. Gradually he stopped moving, and soon he couldn't swallow. His vision and hearing vanished. The boy was finally diagnosed at age two, and he died at four. At that time, Tay-Sachs wasn't commonly known as a "Jewish genetic disease"; its young victims just faded into the background of the many children dying of infection. But the rabbi and his wife knew enough about genetics to understand the recessive pattern of inheritance, implicating them both as carriers. They knew the disease could strike another child.

When the rabbi's wife became pregnant again, the couple refused an enzyme test that could have revealed if the child-to-be had inherited the disease. The only option was abortion, which Orthodox Judaism forbids. So they waited, hoped, and had the baby. After a few months of watching their new daughter decline, they sadly realized that she had Tay-Sachs disease, too.

The Eksteins then had a healthy child, but when their third

child with Tay-Sachs disease was born in 1983, the rabbi decided to act. He started talking to his neighbors, but quickly found that the stigma was huge—the community *did* know about the disease. People hid their sick kids or sent them to institutions. Parents didn't speak of the horror, lest they spoil the marriage prospects of their healthy offspring. Meanwhile, the rabbi and his wife had a fourth child with Tay-Sachs.

Rabbi Ekstein became a community activist before anyone had coined the term, knocking on doors, trying to get people to recognize Tay-Sachs disease. In the beginning, people either didn't understand the distraught man, weren't interested because it didn't affect them, or didn't want to talk about it because it did. But as the rabbi kept reaching out, slowly people started to listen.

As he taught himself genetics, Rabbi Ekstein realized that carrier testing could prevent the births of children with the horrific disease by preventing marriages between carriers—which would take abortion out of the equation. But the other rabbis and community leaders feared that carrier testing would stigmatize people. So Rabbi Ekstein devised a confidential testing program to identify people carrying certain recessive mutations in his community. He named it Dor Yeshorim, Hebrew for "upright generation."

In 1983, Dor Yeshorim began in the Borough Park section of Brooklyn. Young people took tests at school, before they were in serious relationships or matches had been made. Today the program is still going strong, with more than a dozen other diseases added to the roster. It uses accredited medical testing labs and numerical identifiers on blood samples for confidentiality. The Dor Yeshorim database records minimal information—test results, birth dates, and phone numbers. Couples call to learn if both partners carry the same disease. If only one person is a carrier, they're not told, because they couldn't have an affected child. Genetic counseling is provided when both partners carry a mutation in the same gene and therefore face a one-in-four chance of conceiving a child who will be sick, each time.

Dor Yeshorim encountered great opposition at the outset. Only 45 people were tested for Tay-Sachs disease that first year, 1983, and most of them were friends of Rabbi Ekstein. The next year 175 people were tested, then the year after, as word spread, 750. By 2010 more than 300,000 young people had taken carrier tests for an ever-expanding list of Jewish genetic diseases through Dor Yeshorim, and similar programs operate now throughout the world.

Single-gene diseases are rare. For every hundred or so couples who take batteries of Jewish genetic disease tests, one couple is found to be "incompatible." People respond to this news in all ways—they adopt, use donor eggs or sperm, screen embryos, don't have children, or break up. Since Dor Yeshorim started, no children have been born in the Orthodox Jewish community in Borough Park with any of the tested-for diseases. Tay-Sachs is virtually gone from the Jewish community. Brooklyn's Kingsbrook Jewish Medical Center circa 1970 had sixteen beds for Tay-Sachs that were always filled. The facility hasn't had a case since 1996. Overall, genetic testing programs have prevented the births of thousands of children who would have been desperately ill.

The seeds for Dor Yeshorim lay in a program that tested for Tay-Sachs disease carriers on East Coast college campuses with large Jewish populations in the mid-1970s. (Since my college fit the bill, I took the test then.) Michael Kaback, a retired professor of pediatrics and reproductive medicine from the University of California, San Diego, led the Tay-Sachs disease carrier testing program in the 1970s and he's tracked cases ever since. The devastating disease isn't entirely gone. "From 2002 to 2003, there were zero new cases in the U.S. and Canada, but from 2005 to 2006, there were two to four cases a year, which could have reflected the influx of Russian Jews," Dr. Kaback says. Today, ten to twelve Tay-Sachs babies are born worldwide every year, but they are either to non-Jews or to people who are Ashkenazi Jews but don't know it. Other pockets of people—including the Old Order Amish in Pennsylvania, the Cajuns, French Canadians, and Irish—still occasionally have children with the disease, thanks to founder effects and restricted marriages.

Dor Yeshorim was very much ahead of its time. Genetic testing is no longer a stigma in the Orthodox community or in other communities, because so many people do it. Thanks to direct-to-consumer tests available on the Web, requiring only a credit card and a scraped or spat DNA sample, more and more people are discovering that they carry *something.* In fact, each of us has about 140 recessive alleles that, if paired with a second mutation in the same gene from a partner, would harm the health of a child. Before the human genome was sequenced, this "genetic load" was thought to be only 7 or 8 potentially dangerous recessive alleles per person. As the cost of sequencing DNA has fallen and the results of the human genome project have been translated into tests, a panel of nearly five hundred genetic tests will soon be available—to anyone.

Labeling an inherited disease as "Jewish," such as Tay-Sachs or Canavan disease, or "black," such as sickle cell disease, has hampered diagnoses in patients of other backgrounds. DNA is merely a molecule. It is replicated and sent into sperm or egg without considering the religious preferences or skin color of the person. The genes behind Canavan disease and sickle cell disease are in everyone, and can mutate in anyone, because that is what DNA sequences do—they change. Such spontaneous mutations arise from the nuances of nucleic acid chemistry. And so the stereotyping of a genetic disease, thinking that it can affect only certain groups, misses people.

Of the first fifty-two cases of Canavan disease described in the medical literature, thirty-one were Jewish—a majority, but not a sweeping one. Since that time, more than half of the seventy or so known mutations that cause the disease have been found in non-Jewish families. "Neurologists see a child and say it is impossible, it can't be Canavan because he or she isn't Jewish. There are probably many undiagnosed children. Neurologists also don't suggest screening for rare mutations. It's expensive, and people don't bother because the probability is low," says Paola Leone. The cost of the testing escalates with the rarity of the condition; this is why Corey was first tested for albinism, and then common variants of retinitis pigmentosa, before his doctors went on to the rarer LCA.

Lana Swancey, born on April 30, 2001, encountered the problem of the ethnically stereotyped Canavan disease. Her mother, Michelle, had had an uneventful pregnancy. Prenatal testing for common chromosomal conditions such as Down syndrome had indicated a "genetically correct female," but amniocentesis doesn't screen for rare single-gene disorders, unless there is a family history of a specific disease and an additional test tacked on. Because this wasn't the case for the Swanceys, Lana's rapid downward course, after a promising start, was a shock.

"By the time Lana was two weeks old she started having ear infections. By four weeks she was already trying to turn over, and we had to put the sides up on the crib. Then she started going downhill. She couldn't hold her head up anymore," recalls Michelle. She told the pediatrician her concerns at Lana's two-month healthy-baby checkup. He looked at her quizzically. "You're just comparing her to your son, and she's slower. That's all."

"But I *know* something's wrong," Michelle persisted. "Maybe her turning over so early was a fluke, but she was starting to do things, and now she's not."

"Just bring her back in a few months, but I really think she's fine," the doctor said.

By September, Lana hadn't settled down like her five-year-old brother, Coty, had at that age. "I remember sitting on my living room floor rocking a screaming baby girl and seeing the twin towers collapsing on television. I thought life was more than I could handle. My heart was so heavy and sad that I wasn't sure who was crying harder, me or my precious baby girl," Michelle recalls.

Lana's eyes were no longer tracking. She still wasn't holding her head up or rolling over, and her arms and legs were starting to spasm. So Michelle went back to the doctor. This time he paid attention.

"You know, maybe you're right. I think she has cerebral palsy."

"What? But that doesn't make sense. Why didn't we see it at birth? She was doing fine, and now she's not!" Michelle blurted.

When the little girl started to fail faster, she was finally admitted to Richland Memorial Hospital in the Swanceys' hometown of Columbia, South Carolina. She had so many MRIs, neurological and eye exams, gastrointestinal probes, consultations with specialists, and blood tests that Michelle thought her daughter wouldn't survive the assaults. "Dr. Caughman Taylor, head of pediatrics, had only read about Canavan and thought Lana was presenting many of the traits. However, because we weren't Jewish, he thought it was unlikely and didn't mention Canavan to us at that point." Michelle's consent form for amniocentesis had indicated the family wasn't Jewish. Dr. Taylor took Lana's blood and urine anyway about halfway through her two-week hospitalization to test for Canavan. "The day before we left for Duke University for further tests, the urine came back with a high NAA. That's when they knew it was Canavan disease, but felt we needed the skin biopsy to verify it," Michelle remembers. Before they were discharged from the hospital, a geneticist stopped in.

"Good news, we've got it!" he proudly announced. Genetic testing on Lana's blood sample had revealed two unusual, different mutations in Michelle and her husband, Gary, who were of Native American and German ancestry. Lana had two different errors in the same gene, like Corey.

"Thank goodness, now we can treat it!" said Michelle with great relief.

The geneticist looked at her strangely. "Have you ever *heard* of Canavan?"

Michelle and Gary shook their heads.

"It's fatal. She might live to be three, and if she does, she will be completely vegetative. She might not make it past a year."

One look at Michelle's stricken face and Dr. Taylor asked the geneticist to leave the room. It was exactly the same scenario the young Jim Wilson experienced when he excitedly told the mother of his patient Edmund that he'd identified the mutation behind the boy's Lesch-Nyhan syndrome. Without a treatment, the finding meant nothing to Michelle.

The shocked mother was now crying. "I'm supposed to take this child and continue to love her, knowing she's going to die? Tell me how to do that!"

Dr. Taylor fortunately had a better bedside manner than the banished geneticist. "Michelle, there are treatments coming, just hold on. See what they say at Duke. As long as she's alive, there's hope."

An ambulance took the family, including Michelle's brother, to Duke University, where little Lana endured the tests all over again, plus a painful muscle biopsy. When it was time to see the Duke geneticist, Michelle couldn't bear it and sent Gary and her brother instead.

"That geneticist said there was a program run by Dr. Paola Leone, but it was closed due to lack of funding, and we should get in touch with her. But don't go chasing moonbeams, there's really nothing out there. I thought, if there's hope, we'll go," recalls Michelle.

The Duke team confirmed the diagnosis in late November, nearly six months after Michelle first reported something amiss to Lana's physician. Since the Canavan gene therapy trial had been going on in the summer of 2001, Lana had lost precious time because physicians who knew her disease only from textbooks thought it couldn't strike a child who wasn't Jewish. Misunderstanding extended to relatives, too. "They said, 'Oh no, we don't have *that*, it's never been in the family before.' My husband's family doesn't get that it's not just my side of the family, somebody passed it down their side, too," recalls Michelle. Friends and relatives grilled her about whether she did anything while pregnant that could have caused Lana's disease. "So I explained that we all have bad genes and it's just when you hook up with someone who has the wrong other gene . . ." She grows silent, then smiles sadly and continues. "My husband, Gary, and I were meant for each other."

Michelle also feels very strongly that she and Gary were put on this Earth to care for their very special daughter. "Do I wish she didn't have Canavan disease? Yes, but I wouldn't change who she is."

14

THE PATENT PREDICAMENT

LINDSAY KARLIN'S GENE THERAPY WAS POSSIBLE BEcause researchers knew exactly which sequence of DNA to deliver to her cells. In the days before the entire human genome sequence was in hand, identifying a disease-causing gene took years. Still, the 1990s was a time of accelerated gene discovery: for Duchenne muscular dystrophy, cystic fibrosis, and Huntington disease in short order. Among them, although not as celebrated, was Canavan disease, but money matters stalled the translation of that academic discovery into a widely available diagnostic test. The Canavan patent nightmare set a negative precedent, showing future researchers and the families who helped to develop gene therapy what *not* to do.

The Canavan gene patent conflict began in 1981, when Daniel and Debbie Greenberg, who lived in Homewood, a suburb of Chicago, had Jonathon. The baby was diagnosed when he was nine months old. Then lightning struck twice and Jonathon's little sister, Amy, was born with Canavan disease too. To help their children, the Greenbergs founded the Chicago chapter of the National Tay-Sachs and Allied Diseases Association. At a meeting of the organization they met Reuben Matalon, a pediatrician at the University of Illinois in Chicago who worked on inborn errors of metabolism. The Greenbergs begged Dr. Matalon to develop a prenatal test for Canavan disease so that couples could avoid having children as sick as their own.

In 1987, after Jonathon and Amy had passed away, the Greenbergs gave the children's brains to Dr. Matalon, to aid in his gene search. They also started a Canavan family registry and convinced more than 160 families to donate tissues for Matalon's research. But according to bioethicists writing about the Greenbergs, the families did *not* give informed consent. They had little idea of what could, and would, be done with the information obtained from their children's blood, urine, and brains. If the Canavan families could have seen what lay ahead, they might have done what Sharon Terry and Augusto Odone would later do—share in the scientific process, getting their names on the key papers and patents.

With support from the families, Matalon set out on the search for the Canavan gene. The first step was figuring out the metabolic missteps, which was his specialty. By 1988, using urine and blood donated from two families, he had discovered the excess NAA and deficient ASPA that are the hallmarks of Canavan disease. He also confirmed the lack of ASPA in skin cells. His invention of the urine test for elevated NAA was a milestone, because it replaced having a biopsy to detect brain sponginess. (Peeing into a cup is preferable to having a hole bored into one's skull.)

Matalon knew a blood test wasn't possible, because the missing enzyme isn't normally in blood, but he initially thought prenatal diagnosis would be possible using amniocentesis to check the fetuses of carrier parents. Excess NAA and deficient ASPA in the amniotic fluid would indicate the disease. In late 1991, he and three colleagues were about to publish a paper on thirteen such families who'd had a Canavan test via amniocentesis and their babies-to-be declared free of the disease when they heard that Molly Green, born in Arlington, Virginia, in June 1991, was starting to show signs of Canavan disease. The paper appeared early in the new year with an ominous postscript noting the misdiagnosis. To make matters worse, the National Tay-Sachs and Allied Diseases Association had just released thousands of postcards with Molly as their poster girl for successful prenatal diagnosis of Canavan disease. Then the prenatal all-clear turned out to be wrong in three other of the thirteen families. Two of the four families sued and settled. Molly Green died before her first birthday. The researchers learned, by these tragedies, that detecting too little enzyme or too much NAA isn't sufficient to diagnose Canavan disease. Identifying mutations, the abnormal DNA, is imperative.

Finding the Canavan gene took five years, and the discovery is nearly always credited to Matalon, who was by then the director of the research institute of Miami Children's Hospital. In a scientific paper, the names in the middle of the author list include the graduate students and postdoctoral researchers who are low on the academic totem pole, but who do the grunt work. Guangping Gao, whose name was second in the four-author paper identifying the Canavan gene in 1993 in *Nature Genetics* (Matalon's was first), will never forget the day of the discovery. "It was August twenty-fourth, 1992, a Sunday morning. My mentor Rajinder Kaul was there, and we realized we had the gene sequence. But Hurricane Andrew had just struck. There we were in the lab, but all the power was out, and trees were down everywhere."

The lab's celebration when the *Nature Genetics* article was published was a turning point in Gao's career. "There were several patients there, and I remember meeting a nicely dressed young boy, with a tie on, sitting in his special chair. But he couldn't raise his head. I was so moved, I just blurted, *'I had his DNA in my hands!'* which was quoted in *The Miami Herald*." Just as little Lindsay Karlin would instantly inspire Paola Leone to dedicate her career to Canavan disease two years later, this boy had a lasting effect on the young Dr. Gao. "There are so many parents and kids in these hopeless situations. I wanted to figure out a way to fix that. In that moment, meeting that boy, I determined that gene therapy was the only way to solve it."

After earning his PhD with the Canavan gene discovery, Gao did postdoctoral work with Jim Wilson at the University of Pennsylvania, where he figured out exactly how adenoviruses killed Jesse Gelsinger by sending his immune system into overdrive. Today Gao is an adeno-associated virus (AAV) guru, working on dozens of varieties of the viruses.

Having the Canavan gene in hand made accurate carrier testing possible, and the Canavan families hoped that such testing would rid the world of the disease, as happened with Tay-Sachs. By 1996, the New York City–based Canavan Foundation was offering free testing. But a huge problem was looming. In 1997, Miami Children's Hospital and the researchers, including Gao, received patent protection for the Canavan gene. That meant that the patent holders could control the gene's use as the basis of a genetic test, perhaps setting high prices or limiting availability. When the patent was issued, two mothers of children with Canavan disease, Judith Tsipis, PhD, and Orren Alperstein Gelblum, were working to designate Canavan carrier testing as the standard of care for the entire Ashkenazi population—not just those that the disease had touched. They were ready for a fight.

Judith Tsipis's son Andreas was born in 1975, and by the mid-1990s he was one of the oldest patients alive with Canavan. Today Dr. Tsipis heads the genetic counseling program at Brandeis University, in Boston. Orren Gelblum's daughter Morgan was born with Canavan's in 1990. Gelblum has a degree in marketing, but at the time the gene patent was issued she was caring full-time for her severely disabled child. She and her husband, Seth, and their children lived in Manhattan. An article about Morgan in the April 14, 1996, *New York Times* had discussed upcoming free carrier testing for Canavan disease at Mount Sinai Medical Center.

Andreas and Morgan passed away in 1997, but that didn't stop their mothers from fiercely fighting the effects of the gene patent—a battle that reverberates today for other genes. Tsipis recalls their strategy. "We were two mothers, but we had the National Tay-Sachs and Allied Diseases Association and the Canavan Foundation behind us. We started the Northeast Regional Genetics Group, and they approved a statement in June 1997 recommending carrier screening for any Ashkenazi couple. We took the statement to ACOG, the American College of Obstetricians and Gynecologists, which published its statement in early November 1998." With that endorsement, Tsipis and Gelblum launched educational programs in the population at large. "Until then only families of affected children knew about carrier testing. We were very excited and ready to go, talking about carrier screening and the number of labs doing the testing," recalls Tsipis. Expanding the testing could prevent the tragic surprise of a recessive disease lurking quietly in a family's genetic background suddenly showing up in a child.

Widespread testing was a great idea until the patent was issued. "Within five weeks, all the testing labs got letters from Miami Children's Hospital ordering that they cease and desist," Tsipis remembers, her long-suppressed anger resurfacing. "And it became very difficult. They imposed many hurdles to what we really wanted, which was open access, affordable screening for Canavan disease."

The Canavan Foundation responded immediately to the hospital's letter with a one-page statement in *The Miami Herald*, headlined "Miami Children's Hospital Proclaims 'We're Here for the Children.' Unfortunately, Not All of Them." Beneath the headline was a photo of the beautiful, smiling Morgan, above "1990 to 1997" in small letters. Following a review of the facts, the announcement ended with "Please do not let any more children suffer needlessly, and please do not let any more parents' hearts be broken."

Miami Children's Hospital forced health care providers to charge $25 per test, and restricted the number of participating labs and the number of tests that they could perform. The chief financial officer of the hospital sent a letter dated November 12, 1998, days after release of the ACOG statement, to a clinic that had been offering the test, asserting that it must license it from MCH: "We intend to enforce vigorously our intellectual property rights relating to carrier, pregnancy, and patient DNA tests." The reason? To recoup the costs of discovering the gene—Guangping Gao's graduate research project using tissues from deceased children. Matalon pointed out in a news article in *Science* that the Canavan parents had contributed less than $100,000 to the gene hunt, and Miami Children's Hospital, at least $1 million a year.

Some genetic testing labs, wanting to avoid controversy, stopped testing for Canavan disease. The parents who had given their children's urine, blood, skin, and brains to Reuben Matalon's lab, in

Illinois and then in Florida, and had had no idea that anyone would profit from the knowledge gained, felt deceived. Their plight attracted attention. Students in the 2002 class of the Chicago-Kent College of Law, working with professors Lori Andrews, Ed Kraus, and Laurie Leader, filed a pro bono lawsuit, *Greenberg et al. v. Miami Children's Hospital Research Institute*, on October 30, 2000, in U.S. District Court in Chicago. The plaintiffs included Daniel Greenberg, Judith Tsipis, and other parents; the Canavan Foundation; Dor Yeshorim; and the National Tay-Sachs and Allied Diseases Association. The defendants were Miami Children's Hospital and the authors of the *Nature Genetics* paper, including those sandwiched between the mentors, such as Gao. The main legal challenge was not patenting the gene, but using tissue donated for the public good for financial gain. The plaintiffs requested $75,000 in royalties and to block commercialization of the test. The case was settled on August 6, 2003, because, Tsipis says, the plaintiffs ran out of money.

The settlement was a compromise of sorts. Labs could still charge for a Canavan test, but researchers didn't have to pay royalties to MCH for use of the gene in gene therapy, mouse experiments, or other research. Said Matalon at the settlement announcement: "This is a disease where collaboration between investigators of the disease and families of affected children remains critical for advancing knowledge, for prevention and hopefully, for helping affected children." Many in the Canavan community felt his words were too little, too late.

Greenberg et al. v. Miami Children's Hospital et al., although it didn't go to trial, catapulted Canavan disease into the bioethics journals and textbooks. It joined the famous cases of Henrietta Lacks and her cervical cancer ("HeLa") cells, and John Moore and his celebrated spleen. What the three cases shared is exploitation of body parts.

HeLa cells, used and tested everywhere, from the cosmos to the depths of nuclear reactors, came from the cancer-ridden cervix of their owner, a poor, uneducated African-American woman, in 1951. Henrietta Lacks was being treated in the so-called black wards of Johns Hopkins Hospital in Baltimore, and her tumor cells were sampled, grown, and eventually sent to labs all over the world, without her or her family's consent or even knowledge.

The situation for John Moore was quite different from that of Henrietta Lacks. In 1976, the thirty-one-year-old was working long days surveying the Alaskan pipeline. He felt sick, but attributed the bruises, bleeding gums, and expanding midsection to his stressful job. A local doctor recognized the symptoms of leukemia, and Moore indeed had a rare form of the blood cancer, hairy cell leukemia. His doctor referred Moore to David Golde, a specialist at UCLA who removed Moore's spleen, which weighed as much as a large cat. Thinking his disembodied part of no further use, Moore gave permission for the hospital to "dispose of any severed tissue or member by cremation." But Dr. Golde had discovered that the spleen produced a rare and potentially valuable protein, which he'd suspected from the unusual way that Moore responded to certain drugs. The doctor asked Moore to periodically return to UCLA for "follow-ups," even after his patient moved to Seattle.

Moore complied until 1983, when he received a new informed consent form that mentioned "any potential product which might be developed from the blood and/or bone marrow obtained from me." Suspicious, he refused to sign. Dr. Golde then pursued him, even visiting him in Seattle, asking repeatedly for his signature. Growing even more suspicious, Moore showed the consent form to an attorney, who discovered that Dr. Golde had been using Moore's "donations" to nurture a cell line that would churn out valuable proteins, and that he already had a relationship with a biotech firm to develop a product. And in January of that year, the University of California had filed a patent for a "unique T-lymphocyte line and products derived therefrom"—John Moore's spleen cells. The patent was issued in 1984.

When Moore learned that his spleen had been turned into a profitable cell line dubbed "Mo," he said he felt like a "piece of meat." In 1990 he sued the Regents of the University of California on thirteen counts, including lack of informed consent, breach of fiduciary duty for use of his tissues, and failure to disclose personal interest. But the California Supreme Court ruled against him, finding that removed cells are not the equivalent nor the product of a person. After John Moore's thorny case, no one would touch tissue ownership claims for a long time.

The issue of ownership of one's body parts, if said parts are no longer of their bodies, has exploded in the post-genome-sequencing era as people routinely send off DNA samples to companies that offer genetic tests on the Internet. The claim to ownership of two breast-cancer susceptibility genes, *BRCA1* and *BRCA2*, by Myriad Genetics, of Salt Lake City, has reawakened the issues introduced in the Canavan case. Myriad's hold on the breast-cancer genes has kept testing costs high—in excess of $3,000 to sequence the entire gene—and restricted research. A number of genetics organizations filed a class-action lawsuit against the company in 2010, and they won, but their victory was overturned. Myriad Genetics still reigns over the two breast-cancer genes, controlling testing.

People outraged by the breast cancer patents today might not realize that the action traces its roots to the tiny Canavan community. Says Tsipis, "Myriad is the one that will really set the precedent, because breast cancer affects everyone. We were the early ones, the small fry. But we were very, very vocal."

15

CHASING MOONBEAMS

THE FIRST GENE THERAPY EXPERIMENT FOR CANAVAN disease occurred, happily, before the patent problem, but created its own firestorm of media attention.

When Lindsay Karlin and her family met the researchers at Yale in May 1995, Matt During held faculty appointments there and at the University of Auckland in New Zealand, where he was a citizen. Roger Karlin and Matt began speaking daily, both growing excited about the possibility of gene therapy. The discussions went on for months, as the Karlins and the other Connecticut Canavan family, the Mushins, started fund-raising.

Matt and Paola planned to deliver healthy Canavan genes to the children's brains in tiny fatty bubbles called liposomes. With the patient under general anesthesia, a neurosurgeon would drill a hole into the skull directly over one of the four spaces in the brain called ventricles. Continuous with the spinal cord, these pockets contain cerebrospinal fluid, the material that is sampled in a spinal tap. A thin catheter snaked through the hole would deliver the liposomes loaded with the genes, which Matt and Paola hoped would make their way to key neurons in the girls' brains. There, the cells' outer membranes would envelop the liposomes and draw them inside, where the tiny bubbles would release the genes that held the instructions for the missing enzyme. If, after some time passed, brain scans showed myelin beginning to coat neurons, the researchers would know they'd struck gold.

On October 29, 1995, *The New York Times* hailed the coming gene therapy clinical trial at Yale. Two words *not* in the article were *New Zealand*.

"Matt told me he'd be moving his lab to New Zealand and directing gene therapy, and that he might get approval there in a few months, where here it would take years," Roger recalls. The two families planned their trip to New Zealand.

In early March 1996, Lindsay Karlin and Alyssa Mushin had gene therapy, but it wasn't easy, Helene Karlin says. "People were accusing us of going to New Zealand to bypass the FDA. We did think it would be quicker, but we didn't bypass any regulations." Roger picks up the narrative, defending their New Zealand experience. "We had to meet with committees, medical and ethical. The ethics committee people were from all walks of life. Doctors, researchers, nurses, even people from the native Maori community came. The Maori didn't believe in using modified human DNA, and they wanted to know that what was being put into the girls wasn't that. Because the gene therapy was synthetic, it was okay."

But the gene therapy still wasn't officially approved by the time Roger, Helene, and their three daughters made the long trip. On their layover in Los Angeles, Helene made a few phone calls and learned that the ethics committees had given the go-ahead, but the heads of other hospitals had intervened. They finished the journey, then spent a week touring as the committees debated what should be done with their daughter. "On the eighth day we finally heard it was approved, so we went to Auckland and got it done," recalls Roger. The okay came from a panel assembled by the New Zealand Health Research Council.

The family faced opposite reactions on opposite sides of the planet. Molly Karlin, who was thirteen at the time, remembers the phone call from Matt During while they were at a hotel in a rain forest, telling them the approval had finally gone through. "Lindsay had the gene therapy the second week we were in New Zealand, and it was all over the papers. We were celebrities in Auckland.

People stopped us in the street. A Maori priest came to the hospital and blessed Lindsay and Alyssa," Molly recalls.

Matt announced the gene therapy in a news release March 6, trying to temper expectations. "The best we can hope for is that the procedure is safe; anything over and above that will be a bonus." A few days later, *Science* responded with an uncharacteristically inaccurate news story, "New Zealand's Leap into Gene Therapy," that quoted an NIH Recombinant DNA Advisory Committee (RAC) member accusing During of intentionally ignoring rules that other scientists obey, and called Canavan disease the illness that "figured in the movie *Lorenzo's Oil*"—which was adrenoleukodystrophy. The article called During a visiting professor at Yale, when in fact he was an associate professor of surgery and medicine and had been director of a gene therapy lab for eight years, which During pointed out in a letter to *Science* published on April 26. The article also got the dates wrong for when During contacted regulatory agencies, and juxtaposed quotes to imply that Matt During was fleeing the United States to rush a risky gene therapy experiment into being elsewhere. In June *The Lancet* published a news article, "Gene Trial Causes Ethical Storm in New Zealand," which was a claim very different from the Karlins' experience in the days following their daughter's surgery.

The *Science* and *Lancet* stories infiltrated the mainstream media, and the resulting criticism would haunt Matt During for years. If he and Paola Leone wanted to expand the clinical trials for Canavan disease—which they certainly did, and they already had dozens of families pleading for a place in the queue—they'd have to counter their images as renegades.

The Karlins helped. They became regular speakers at meetings of the FDA, the RAC, and institutional review boards. "They didn't want us at any of these meetings, but the public is allowed, so we'd show up anyway," Helene recalls with a chuckle. "We kept saying that these children follow an invariably fatal, degenerative course. All the studies on safety seem to be promising, there's some indication of efficacy, and therefore these kids have an ethical right to get the treatment. But even if hospital administrators were leaning to approve, lawyers would jump in and say NO! They didn't want to take any risk. But it was better than waiting for no treatment and a funeral."

While Matt, Paola, and the Karlins tried to blunt the protests over the gene therapy done in New Zealand, they also had an eye on the two young patients. And what they saw was encouraging.

"Before, Lindsay never made eye contact. She'd smile at the wall. The day after the gene therapy, she looked at me and she smiled," remembers Helene, a high point that Roger witnessed too. Alyssa did the same. Both girls also seemed to acquire some head control, becoming a little less floppy.

More objective measures were promising too. Magnetic resonance spectroscopy of Lindsay's brain over the next year showed a sustained drop in NAA level, while Alyssa's dropped and went back up after nine months. Alyssa had evidence of the enzyme ASPA a year later, while Lindsay did not (though During says it could have been missed in her spinal tap, which is a sampling of the fluid). Although Alyssa would never hold still long enough to have her vision checked, Lindsay's vision greatly improved, a finding that other parents would report in future trials. The optic nerve pathway to the brain is short and it myelinates the fastest. Most telling, though, were the serial MRI brain scans that showcased the changes in Lindsay's white matter: myelin was appearing in the structures closest to where the genes entered.

It was progress, but incremental. Slowly, Lindsay began to meet developmental milestones. One month after gene therapy, she could bear weight on her legs if someone lifted her from the arms, and turn from her side to her back. "I remember sitting with her and the physical therapist, and she was able to get in a prone position and hold her head up for a few minutes. That was unheard of. She couldn't do this before the gene therapy at all," says Molly. By thirty-one months, Lindsay could reach out with both hands and had attained the social and language skills of an eighteen-to-twenty-four-month-old. Alyssa had some neurological improvements, but hit a plateau after a year. Eventually Lindsay's progress leveled off too.

The researchers knew they had to test the gene therapy on more children, preferably younger ones, as Lindsay had done better than Alyssa. Matt and Paola were also eager to improve the method of gene delivery. Next, perhaps back at Yale, they'd send liposomes carrying a doubled gene dose into a plastic pouch, called an Ommaya reservoir, placed underneath the scalp. From there, catheters would distribute the liposomes to the surrounding brain tissue and hopefully reach more of the places that needed ASPA.

Canavan parents anxiously followed Lindsay's and Alyssa's progress. A third couple, Richard and Jordana Sontag, had the financial resources to get things moving. Their son Jacob had been born on February 24, 1996. The Sontags provided funds to continue the clinical trial in New Zealand, but the *Science* and *Lancet* articles had done their damage. "The gene therapy committee there categorically refused to proceed with the project, claiming that the original approval was granted on a 'compassionate use' basis and would require going back to square one for regulatory approval," Paola remembers. With that pronouncement, the Auckland experiment was over. The children did, however, still have Paola, who returned to Yale while Matt During remained in New Zealand.

Yale University and the parents funded a gene therapy trial that was approved in 1998. The timing was good. Miami Children's Hospital hadn't begun to enforce its patent on the Canavan gene, and it was before Jesse Gelsinger died of his gene therapy in Philadelphia. In this second gene therapy trial for Canavan disease, Jacob Sontag, Lindsay Karlin, and fourteen other kids each received 600 billion to 900 billion viruses through six catheters leading into their brains from the implanted reservoirs. One of the children was Max Randell, an adorable redhead with bright blue eyes, then eleven months old.

Ilyce and Mike Randell, who live near Chicago, had been in the world of Canavan disease for only six months. Like the Karlins, the Randells had been told, "Don't get attached to your baby, and look for a nursing home." A diagnosis as grim as Canavan hits siblings and grandparents, too, and Max's Grandma Peggy remembers their initial panic and anguish well. "Everyone said it was horrible, vegetative. So my sister, a neurologist, went to three medical libraries, to look for a photo of a child with Canavan who was smiling, showing emotion. She found one of a boy with a beautiful smile. Ilyce put it on the fridge, and it gave everyone hope." She stops to collect herself, and her voice grows lower. "Max is now that picture." Soon after Max's diagnosis, the family drove five hours south to visit a little girl with Canavan's, to see what her life was like. Meanwhile, Max's aunt, the neurologist, called Paola Leone. Like the Haas family, the Randells had the incredible luck of finding a gene therapy trial already in the works. Both Corey and Max were the youngest participants in their trials.

With all of the regulatory red tape, it seems a miracle that the Yale clinical trial of 1998 took place at all. The FDA, RAC, and Yale Biological Safety and Human Investigation committees had met often throughout 1996 through 1998 to discuss the proposed clinical trial. Progress toward approval stalled as the problems began to seem insurmountable: the prospect of drilling holes in children's heads and injecting genes into their brains and the one-in-twenty risk of infection from implanting the reservoir, against the backdrop of gene therapy's uninspiring track record of two-hundred-plus trials with no definitive, sustained disease correction.

Practical matters stalled progress. "It seemed like every time there was an ethics committee meeting, some board member would be on vacation. I remember thinking, 'My sister's life is on the line and we're waiting for someone to get back from Martha's Vineyard?'" recalls Molly Karlin.

The summer of 1997 was devastating to the tiny Canavan disease gene therapy community. Morgan Gelblum died, at seven years old, weighing just thirty pounds. While the Yale committee members were on hiatus, Jacob Sontag was declining. The changes were subtle, more obvious to a parent than to a doctor's periodic exam or peeks at test results or scans. Jordana knew that Jacob was taking slightly longer to recognize her, and one of his eyes was turning a bit—he was losing muscle tone more quickly now. When the Yale biosafety committee again couldn't reach a decision in September, Richard and Jordana packed up little Jacob, drove to Yale from their New York City suburb, and shoved him in the face of a key administrator, to the astonishment of a watching gaggle of new medical students. Their ploy seemed to work, because Yale gave the trial the go-ahead in October at its next Biological Safety and Human Investigation committee meeting, and the government came on board in December 1997. The various committee members finally agreed that the therapy could only help the Canavan kids.

Matt and Paola flipped a coin to decide who would go first, Lindsay or Jacob. The boy won. On Friday, January 2, 1998, early in the morning, the surgeon Charles Duncan inserted the egg-shaped reservoir, which bulged from the boy's forehead. The gene infusion was set for Monday, but Jacob spiked a fever from a respiratory infection, and then had a seizure that landed him in the intensive care unit. He wasn't well enough to receive the gene therapy until January 22. By then, the procedure itself was anticlimactic. In less than five minutes, Dr. Duncan squirted the genes into the egg on Jacob's forehead, pumped the syringe a few times, and that was it, like basting a turkey.

After the gene therapy, the Sontags entered the strange world of expectation. Any possible change, no matter how tiny, might signal that the genes had hit their mark and were silently remyelinating the naked neurons. If Jacob touched Jordana's hand, she was ecstatic. Wanting to maximize his chances of improving, the Sontags upped his physical therapy sessions to five times a week, though this added another confounding variable. When they started giving him a drug at night to combat the sleeplessness that had followed the gene therapy, it added another factor. A journalist covering the family's experience for *The New York Times* concluded, after watching the family for months, that yet another variable may have been the most significant—the love and attention that Richard and Jordana lavished on their son. Augusto Odone had similarly wondered about his son, Lorenzo. Had the intense, extended, excellent care obscured any benefits of taking the eponymous oil?

By June, Jacob showed some significant gains and his MRI scans showed some remyelination. Meanwhile, Matt and Paola moved to Philadelphia to direct a gene therapy center at Thomas Jefferson University. The significantly greater lab space and several million dollars of institutional start-up funds made an offer they couldn't refuse.

Lindsay improved for two years following her second gene therapy, her parents claim. Max did well too. "After the first treatment, the first thing we noticed was his improved vision. Within two to three weeks, he started tracking with his eyes, and he got glasses. He became more verbal and his motor skills improved. His vision is still so good that his ophthalmologist only sees him once a year, like any other kid with glasses. She calls him Miracle Max," says Ilyce. Mike believes that the gene therapy is probably what has kept Max alert and communicative, despite his physical limitations, which are great. "I think the gene therapy prevented the vegetative state." Max may have improved—or worsened more slowly—than the other children because he was younger when he had the gene therapy.

Although other children improved and the ultimate conclusion from this clinical trial was that the gene therapy was working, not all of the participants did well, because the procedure itself was invasive and therefore risky. Jacob Schwartz was sixteen months old when he had the reservoir implanted and gene therapy in Philadelphia. A few weeks later, a two-hour seizure was the first sign that he had a raging brain infection. He needed surgery to remove the reservoir and antibiotics for several months in his home city of Toronto. His parents did not return to the United States for follow-up exams, and Matt and Paola couldn't assess the effects, if any, of the new genes bathing his brain.

Paola Leone was constantly trying to improve the gene therapy. The initial pilot study on Lindsay and Alyssa in 1996 used liposomes, as did the clinical trial on them and fourteen others in 1998. By 1999 Paola was developing a viral vector for a new trial, which would replace the liposomes with viruses. The viral route, she thought, would produce a more lasting effect. The new vector was AAV2, the same vector later sent into Corey's eye. Paola would conduct the new clinical trial as director of the Cell and Gene Therapy Center at the University of Medicine and Dentistry of New Jersey, where she is today associate professor of cell biology. But that trial hit the mother of all roadblocks in September 1999—Jesse Gelsinger's death happened just across the river.

After Jesse's death, the NIH halted many gene therapy trials, and allowed very few new ones to start. The Canavan parents had to act, and fast. "We got as many families as we could, and took the kids to Washington, D.C., to lobby in support of gene therapy for Canavan patients," recalls Ilyce Randell.

The ensuing multiyear wait for gene therapy, from 1998 until 2001 or 2002 for some kids, was devastating; progress slowed and stopped. "There'd been stress, from morning to night, for years. Raising money and trying to fund the science, amassing safety data, worrying about efficacy. Then everything was put on hold, with the FDA denying it was on hold. And so we went back and forth to Washington with kids in wheelchairs," Ilyce recalls. Even though the trial was stalled, to keep things on schedule, in case the trial was given the green light, the families traveled to Philadelphia for MRIs, which had to be done on the same machine each time for consistency. The regimen was hectic. "By the time we finished getting the postoperative data for one trial, we were getting presurgical data for the next one," Ilyce says. For many years the family made minivacations out of the trips, but eventually, years after Max's second gene therapy, they stopped trekking to Philly for follow-up.

Two years post-Jesse, the FDA began to allow some gene therapy trials to resume, or begin, because research on making the vectors safer, thanks to Jim Wilson's team and others, had never stopped.

The third try at gene therapy for Canavan disease, using the viral vector that Paola had developed, had two parts. Lindsay, Max, and Jacob Sontag were treated first, in that order, in the summer of 2001. The procedures were done at Jefferson Medical College, using funds that the parents had raised. Then a $1.8 million grant from the NIH covered an additional fifteen kids treated at Cooper Hospital, affiliated with the University of Medicine and Dentistry of New Jersey, and followed up at CHOP. The informed consent document reflected the attention to detail in the post-Gelsinger era: "Although this treatment is popularly referred to as 'gene therapy,' we cannot guarantee that any therapeutic or beneficial result will follow, and it is best to consider this trial a gene transfer experiment." This trial was the "moonbeam" that little Lana Swancey was told not to chase.

In this third trial, the neurosurgeon Andrew Freese placed 900 billion viruses, each harboring a normal Canavan gene, into catheters entering each child's brain through six holes. This time, instead of using a reservoir, the neurosurgeon looked at MRI scans to place the catheters, delivering the therapy directly to where the white matter was melting away in each child's brain. If all went well, the viruses would enter some 180 million of the right cells and safely deposit instructions for making myelin. The target was fewer than 1 percent of all the brain cells, but hopefully it would be enough to make a meaningful difference in quality of life for the children.

The procedure took just under three hours, and required a two-to-four-day hospital stay. The children were then tracked every few months with brain scans, urine tests for NAA, spinal taps for the enzyme, and neurological exams. Again, they improved—a little. A child's making eye contact, cooing at the sight of a sibling, or lifting her head was the equivalent, to a Canavan parent, to Corey's being blinded by the light at the zoo four days after his gene therapy. It worked, the parents insisted. But there were no experimental controls, no Canavan children given sham surgeries to see if the six bored holes themselves tweaked brain activity. Still, parents really couldn't help their optimism. Lana Swancey, the cute little girl with the long, straight brown hair whose diagnosis had been delayed due to her non-Jewishness, had gene therapy the day after her second birthday, in April 2003, and the change was both immediate and dramatic.

"Lana was a cranky, crying baby. She screamed constantly. But the day after the gene therapy, she was smiling and acting like a different child. She was having such a good time they had to move us out of the ICU. She hadn't smiled or laughed in two years," says Michelle. The changes continued. "In the first couple of months she started babbling, tracking, and her movement was not as spastic. She could get up on all fours, turn over, and hold her head up. If only we had had the gene therapy soon after she was born . . ."

Many of the Canavan kids were featured on their local TV stations and in their hometown newspapers, but the national media were impatient. The small changes that thrilled the parents were not the overnight breakthroughs that make good news stories. The Karlins are still bitter over their experience with the television program *60 Minutes*.

In New Zealand and Australia, *60 Minutes* had broadcast an uplifting story about the first Canavan trial, featuring Lindsay and Alyssa. "It was a very well done, scientific human-interest story," recalls Helene Karlin. A few years later, the show wanted to do a follow-up and filmed footage of Lindsay in Philadelphia for the trial at Jefferson. "But by that time, Lindsay looked pretty bad," Helene says. "She'd gone from a cute little baby to a child with a disability. She was bigger and had a major disease," continues Roger. The little girl spent most of her time and energy just trying to keep her eyes open—not exactly scintillating television. "So they shot all this footage at Jefferson, but at the last minute yanked the program. The producer called, upset, and said that her boss had decided not to air it because it was too depressing," Roger recalls.

After the third try, gene therapy for Canavan disease seems to have faded away. A 2006 report in *The Journal of Gene Medicine* concluded that the viral gene transfer was safe, the dose and delivery method okay, there was no serious immune response to the virus in most patients, and a phase 2/3 trial was justified to show efficacy. But such a trial hasn't taken place. The reasons are complex, but center around the severity of Canavan disease and the limited, if lasting, improvement following gene therapy.

Matt During went on to cofound the biotech company Neurologix, where he returned to investigating Parkinson's disease. In a phase 2 trial the company is sponsoring, gene transfer alters the level of a key neurotransmitter as intended by the investigators and has led to a twofold improvement in motor skills.

Paola Leone remains as dedicated as ever to the Canavan kids, but she's moved on to another technology. "Delivering the gene itself isn't going to cure the disease—not with the dramatic brain atrophy and cell loss that happens in the first twenty-four months," she explains. Instead, she's using human embryonic stem cells from the privately held Geron Corp. in Menlo Park, California, that have normal genes and can be coaxed to specialize as oligodendrocytes, the cells that naturally supply the enzyme missing in the disease. "This will target the cells that are affected the most, so I don't have to overexpress the gene in every cell. The stem cells can replace lost cells."

The Canavan saga raises a compelling question: What happens *after* gene therapy? Have the incremental changes justified the risk? The parents think so. Without gene therapy, Lindsay, Jacob, Max, and Lana might be "vegetative," as their parents had been told they would be, or not here at all. Most children with classic Canavan disease die before age ten, usually of aspiration pneumonia or a seizure. These cherished children who have had gene therapy for Canavan disease remain the centers of the universe for their dedicated families.

Lindsay Karlin has stayed about the same since age seven, when her gene therapy stopped. But that's enough for Roger and Helene.

"Gene therapy has taken us off the roller coaster of fighting against time, because we don't have to live in dread every day, thinking, 'What deterioration will we see today?'" says Helene.

Life has gone on around Lindsay, and has included her. When she turned thirteen, her ten-year-old sister, Jolie, whom Helene and Roger adopted, stood up at temple and read from the Torah in place of her older sister, so that Lindsay could be Bat Mitzvahed. Lindsay's Sweet 16 in July 2010 was more low-key—the family went to the movies and out to eat, including birthday cake. Lindsay is rare among Canavan kids in being able to eat pureed foods, rather than being fed through a tube leading into her stomach, and that day, she had pureed birthday cake.

Lindsay goes to school, has frequent therapy sessions, and uses a machine that exercises her limbs at least an hour a day. She loves movies and listening to music on her iPod, especially an opera recital of her sister Samantha singing, and the Beatles. Says Roger, "She's really responsive and sweet. We wish she could do and see more, but she's happy. I ask her if she's happy, and she makes a noise to let me know that she is."

Like the Karlins, Jordana Holovach doesn't regret Jacob's gene therapy for an instant. She and Jacob's father, Richard, divorced. She remarried and now has two daughters, Remi and Hailey, who adore their big brother.

"The benefits of the gene therapy include Jacob's overall health. He barely ever gets sick from the common cold, or the flu. I know in my heart that the gene therapy has kept him from slipping into a vegetative state. He is with us, he engages us, he laughs appropriately at what is funny and responds appropriately when we talk to him," Jordana says. Although compared with other kids there is much that Jacob can't do—such as move or eat—his days are very full. "Jacob attends school twelve months a year. He uses a computer by moving the cursor with his eyes. He attends a program with other special-needs kids that includes games, daily living activities, and music. At home, he has occupational therapy, motor therapy, and massage," Jordana says. And like Lindsay, Jacob loves listening to music on his iPod.

Coty Swancey was five when his sister, Lana, was born. Today he is her biggest advocate. "I've spent most of my life right beside my sister, singing and reading to her or in the hospital holding her hand. I see her fighting every day and still having fun, smiling the whole way through. We travel to Pennsylvania and New Jersey from South Carolina at least once a year for Lana's checkups. My favorite thing about the hospital is that no one looks at Lana like she is strange, they look at her like the beautiful person that she is. Right now Lana is doing just fine, still doing therapy every day, still smiling. As for me and my family, we will be there for her always," Coty blogged when he was fourteen. He worries about the future—not that Lana won't be here, but that he won't be able to care for her.

As I listened to the parents of the Canavan kids share their stories, I sensed that these children are truly not like others with different brain diseases. I recalled asking Amber Salzman if she would have wanted gene therapy for her nephew Oliver, who died at age twelve of adrenoleukodystrophy. She started to cry. "Why would anyone want to prolong the state he was in?"

Maybe Amber's understandable response reflects the fact that in ALD, kids are okay for their first few years. But Canavan kids are never okay, and so any small change toward normalcy is a victory. Yet I didn't quite understand how the parents could be so sure that aware individuals lay trapped in their eerily still children—until I met Max.

I began writing about Max Randell for my human genetics textbook in 2001, when he was three years old. Each edition chronicled his progress, and I always wanted someday to meet him. I finally did on Saturday, October 9, 2010, at the annual fund-raiser for Canavan Research Illinois, the not-for-profit organization that the Randells started. The occasion marked Max's thirteenth birthday.

When I saw Max for the first time, I knew just from looking into his eyes that he was not in any sense "vegetative." But if I'd needed any further convincing, I'd have gotten it minutes later, when Paola Leone arrived and bent down to place her face right before his eyes, so that he could see her.

"Max, it's Auntie Paola!" squealed Grandma Peggy, a cheerful woman with short gray hair jumping up and down in excitement.

"Hi, Maxie," Paola said, brushing his cheek with her fingertips as the boy's face transformed. His brows lifted, his eyes lit up, and he broke out into a smile so big that it simply defied the stripped axons that had robbed him of so many other movements. He gurgled a greeting.

Max continued to beam, tethered to his wheelchair, decked out in a blue suit with a red pin-striped tie, as guests took turns kissing him. I pulled up a chair and sat in front of him to show him the birthday cards I'd brought from students who'd read about him in my textbook. Then Peggy and Max demonstrated his language: one slow blink for "yes" and a widening of his eyes for "no." He didn't have enough muscle control to blink twice.

Two smiling little girls, also with Canavan disease, were wheeled by. They still had some movement, and they were flipping their hands back and forth, up and down, as if trying to fly away. But their hand movements were "decerebrate posturing," the automatic activity of a damaged brain rather than purposeful movement. A dozen young kids ran about waving purple glow sticks, and I couldn't help but wonder what Max thought of it all. He seemed happy, watching them.

I sat at a table with Paola, Mike, Max, and Max's charming little brother Alex, who was eight going on twenty-eight. While the adults ate chicken alfredo (except for the vegetarian Paola) and the kids downed chicken fingers and fries, Mike leaned over Max, pulled up his shirt, and deftly attached a bag of cream-colored stuff to the feeding tube leading to his son's stomach, holding it aloft for the food to go in. At the front of the ballroom, Ilyce, svelte in a shimmering blue gown, was getting ready for her annual thank-you.

Max was finishing his meal when Ilyce, looking across the room at him, began. "Becoming thirteen is a monumental day. Max is a happy and wonderful boy. He was diagnosed March sixth, 1998, when he was four months old." As she recounted the story of the two gene therapies, Max watched her intently, coughing and gurgling a little bit. Then Ilyce could no longer hold in her emotions. "Maxie, you are my treasure," she said to him from across the ballroom as her tears streamed. The boy's face mirrored hers, and suddenly he was sputtering and gasping, reacting viscerally to her words. Mike settled Max down enough to listen to Alex, who got up to speak next. The boy, wearing a suit like his brother, took the microphone and read his prepared speech. The grin that had been perpetually splashed across his face vanished as he talked seriously about the things he likes to do with Max. He ended, "Max, even with Canavan, is the greatest brother ever. I love Max very much."

After the speeches, Mike wheeled Max from the ballroom, maneuvered him out of the chair, and settled him on Peggy's lap, where he lay limply. Peggy gently held his head as Max smiled up at her and said "Aaahhh." It was clear that the two have a special relationship.

"You're first of my five grandchildren, all boys," she said to him, then looked at me. "Can I tell her our games? How about What If Everyone Was in a Wheelchair? or Let's Think of a Color." Peggy thought a minute, then slowly named colors. Max blinked a yes when she named the one he thought she was thinking about. Then Alex popped over for a visit and Peggy drew him into a two-boy bear hug. Max grinned again, as he always does around his little brother. "Max loves when I read him Harry Potter," Alex said. Then he checked the cell phone peeking out of his pocket and was off in a dash, joining the other kids running around.

Meeting Max had a profound effect on me. Like Paola Leone and Guangping Gao, I sensed that these kids are as whole, cognitively, as other kids. Yet at the same time, the physical limitations brought to mind Christopher Pike, the fictitious first captain of the starship *Enterprise* in the original pilot episode of the television series *Star Trek*. Captain Pike had sustained such severe injuries that he was completely unable to move, and communicated by directing his brain waves to turn a light on, with one flash signaling "yes" and two flashes "no." Captain Pike's captors used technology to create an illusionary existence in which he was healthy. We're not there yet.

Until meeting Max, and especially because I write biology textbooks, I had thought of Canavan disease, adrenoleukodystrophy, ornithine transcarbamylase deficiency, and the hundreds of other unpronounceable horrors, the so-called inborn errors of metabolism, as just that—blocked pathways and cycles of biochemical reactions. Paola Leone at times lapses into bio-speak too. At dinner she talked about N-acetyl aspartic acid, the second most abundant amino acid derivative in the brain after glutamate. It is what floods the urine of a little Canavan baby who can't keep her head up or look her parent in the eye.

But Max, Lindsey, Jacob, Lana, and the others are not just the consequences of biochemistry gone awry. They are loving individuals who bring joy as well as heartache to their families. For some of these awe-inspiring parents, what keeps them going, day after day, is the feeling that they were somehow chosen to shepherd these special children through their time here. Sometimes that makes more sense to me than the chemistry behind a mutation and the genetics behind how it is passed on.

By the time that the gene therapy attempts for Canavan disease had quietly ceased, leaving behind children who were better off than they would have been, but still disabled, testing of the gene therapy that would be done on Corey Haas in 2008 was already well under way. The emotionally wrenching Canavan stories had raised the question of whether a *partial* forever fix, a stalling or slowing of disease progression, is a worthwhile goal.

Like Max Randell, Corey was the youngest to receive gene therapy for his condition. The researchers behind both sets of trials, for

Canavan and LCA2, thought that the younger the patient, the more likely a genetic correction would forestall symptoms. But the improvements of the Canavan kids had slowed or even leveled off as time ticked away between gene therapies. Would the children have benefited from more frequent treatments? Will Corey and the others who have had their vision restored need a booster gene therapy? The answer will come from dogs—and that is really where Corey's story starts.

PART VI

COREY'S STORY

We can do anything we want to
do if we stick to it long enough.

—Helen Keller

16

KRISTINA'S DOGS

EARLY ON A SUNNY SATURDAY MORNING IN MIDSUMmer 2010, a small woman with streaming reddish-brown hair struggled into a conference room at a Philadelphia hotel, dragging a large, shaggy dog by its front end. Jean Bennett's head was buried in Mercury's furry neck, and her slight frame and minidress suggested that she was a student—not the professor of ophthalmology who co-led Corey's gene therapy trial. The hotel was a short walk from the Children's Hospital of Philadelphia.

That day was Mercury's first foray out of the laboratory. He'd been born there, bred to inherit LCA2, Corey's disease, and had gene therapy there as a young pup. So far, Mercury's only activity had been running an obstacle course set up in the hospital corridor, created to test his vision. He'd run the course so many times that the researchers had to change it every now and then, because the dog memorized the locations of the objects put in his path.

That Saturday morning Mercury stretched and stumbled at first, like a newborn calf, but within minutes he was completely unfolded and loping about, then barreling beneath the rows of seats, sticking his head up an occasional skirt. He trotted over to the side of the room where Al Maguire, Corey's eye surgeon, was sitting and drinking a mug of coffee. Mercury draped himself across the surgeon's lap and sniffed at the drink. Then Dr. Jean, sitting next to Maguire, her husband, and recovered from her dog transport task, grabbed the fuzzy red leash and brought the dog over to Corey, who was sitting in the front row with his parents, Ethan and Nancy. Seeing the dog, Corey bolted out of his seat, grabbed the leash, and the two took off, running back and forth behind the podium.

The all-day scientific symposium of the 2010 Family Conference for the Foundation for Retinal Research was about to get under way. Out in the hallway lined with exhibits of Braille products and special musical software, families milled about, most with children nearby. Some of the kids had canes, but the older ones, and a few adults, held the leashes of Seeing Eye dogs. The adults were trying to round up the kids, visually impaired as well as their sighted siblings, to send them off on a foundation-sponsored all-day adventure in the historic city.

The conference room grew more crowded after the children left. The buzz of conversation hummed louder, increasingly spiked with little shrieks as excited Facebook friends met in the flesh. The families all had children with LCA, but the disease comes in at least eighteen types, caused by errors in different genes. Gradually, the attendees sorted themselves out by mutation, like students in a dorm gathering into groups based on which high school they'd attended. After greetings, the conversations quickly veered into techno-speak, for these parents are experts in their children's diseases.

"Omigod! You have *GUCY20* too? The chicken disease?"

"No, no, we're *CEP290*. The cat one."

"You're *LRAT*? When did you find out? Are you on our Facebook page?"

"We're *CRB1*. You, too? I think we're lucky. The RPE's there, but the rods are just out of whack. That means gene therapy'll work. Like Corey."

"Yes, she's just very light-sensitive now. Her fundus is still normal, but it's going to atrophy. Unless she has gene therapy."

"*RDH12* was easy to spot, because his retinas are shredded. Like fishnets, the doctor said. But the RPE's okay. We have a chance." The *RDH12* group that formed spontaneously that morning would go on to raise $70,000, which the children presented to Dr. Jean in early spring 2011.

Off to the side stood Troy and Jennifer Stevens, who looked like they were barely out of their teens. They'd traveled from California and seemed more nervous than the other parents. Later that day, between talks, they'd have ten precious minutes with Dr. Ed Stone, director of the Carver testing lab. Dr. Stone had promised to give them their son's results. They'd had to send in blood samples several times. Why was it taking so many months to get a diagnosis? Thoughts of gene therapy couldn't even begin until they knew which mutation in which gene was causing two-year-old Gavin's blindness. Jennifer wasn't very hopeful. "From the beginning I had a feeling the Carver lab wouldn't be able to identify our gene," she told me, on the verge of tears.

Jennifer had learned early on to trust her maternal instincts. "Just after Gavin was born, I knew something was wrong. They handed me a perfect-looking baby, and I knew. It was an overwhelming feeling that I couldn't dismiss," she recalled, wiping away a tear. She thought she was losing her mind. Her friends and relatives chalked her anxiety up to lingering effects of the drugs she'd taken during labor, then to postpartum depression. But the feeling wouldn't go away, and Jennifer insisted to one doctor after another that Gavin couldn't see properly. Finally, months later, the specialists agreed with her. "It was a bittersweet moment. Other people finally understood that something was wrong. But there was sadness, too. The ophthalmologist had just told us our baby was blind. LCA wasn't mentioned at that time. He just said 'retinal dysfunction.'" As Jennifer feared, Dr. Stone would indeed tell her and Troy that their family LCA mutation wasn't among the known ones. Their diagnostic journey wasn't over. (The Stevens family would finally have their mutation identified a full year later, after having their exomes—the protein-specifying part of the genome—sequenced.)

Suddenly the little knots of people all turned in the same direction, and Troy and Jennifer looked too.

"That's them! That's Nancy and Ethan!" a woman shouted as her husband pointed to the couple standing in the front row, watching their son and the giant dog.

Everyone moved toward the Haases. The families took turns photographing one another with Nancy and Ethan, who were stunned at the attention. Although 2006 to 2008 had seemed an eternity to them, in fact Corey had gone so quickly from diagnosis to clinical trial participant that this was their first Foundation for Retinal Research meeting. Some of the other families had been coming for more than a decade. If the families whose children had already lost their sight were jealous, they didn't show it. The room simply radiated joy and wonder.

As Corey and Mercury became more rambunctious, little by little people turned to look, and astonished silence rippled through the large room. Then Betsy Brint, head of the organization with her husband, David, walked to the lectern to start the symposium. But she, too, couldn't help staring at the boy and the dog. "That isn't the blind leading the blind," she paused, her voice wavering, "but the sighted leading the sighted."

Betsy then formally introduced Mercury, a Briard sheepdog. He and his labmates, including his distant cousin Lancelot, the first dog cured of blindness with gene therapy, were up for adoption, she announced, and the attending families had first dibs. Some of the parents started to bounce in their seats, a few waving hands. So Dr. Jean snaked her way in and out of the aisles with a sign-up sheet, and soon each dog had a list of potential loving homes.

Gene therapy for LCA2 was a long time in coming, and would not have been possible at all if it weren't for experiments using animals. When Corey Haas was born, in 2000, Lancelot was just beginning to see, days after his gene therapy. But the story goes back even further, to an astute veterinarian in Sweden who, some say, has never been appropriately recognized for her contribution to Corey's sight.

Kristina Narfström has made a career of studying dogs and cats that have versions of human diseases. Her blind Abyssinian cats have retinitis pigmentosa, and her sightless and seizure-plagued dachshunds have Batten disease, the brain disorder affecting the Milto family in Indiana, for which gene therapy trials in children are under way. Dr. Narfström's work with Briard sheepdogs began in the late 1980s with an intriguing new patient.

By age fourteen, Narfström knew that she would become a veterinarian. She grew up in a family of engineers, with just one dog, Kelly, and no cats. But how she loved that dog! When her family returned home from an extended trip that year, the dog had to be quarantined, and Kristina was very upset. "I didn't want to leave him alone, so I worked at the quarantine. That's when I started to love animals. The vet who came to treat the animals told me I was very suitable for a life as a veterinarian," she recalls.

For a time, Narfström debated whether to become a "human doctor" or a vet. She chose the four-legged animals, but had no inkling that her findings on dogs and cats would one day lead to gene therapy for people. "I get goose pimples when I think about Corey," she says.

Toward the end of vet school, Narfström became fascinated with the visual system, and specialized in veterinary ophthalmology. "I realized that eyes were very important and interesting, and not much was known about them in pets at that time." She got her veterinary degree in 1973, and added a PhD in 1985 working with the sightless Abyssinian cats.

Narfström's discovery of the "dog model" for Corey's disease came, as many scientific strides do, serendipitously—with a dose of Louis Pasteur's "chance favors the prepared mind." In 1988 Narfström was an associate professor in the veterinary school at the Linköping University, where she was seeing patients and doing research. "One day a breeder called, devastated. She had a litter of nine dogs, about six months old. One of the dogs was behaving strangely, walking into things in the dark. There were other indications that all was not right with the litter. The breeder had seen two other vets who'd said, 'Yes, it appears as if this dog is blind, euthanize it.' Instead, she called me. And I asked her to please come in with her dogs immediately."

They were Briard sheepdogs. Narfström had heard of them, but couldn't recall any reports of their suffering from night blindness. So she read up on the shaggy dogs while awaiting the breeder's visit.

The origins of the Briard breed are unclear. Some sources point to Belgium, but the name is French, from *chien berger de brie*, which translates into "the hairy dog of Brie" (referring to the French province, not the runny cheese). The breeder told Narfström that the French dogs actually came from North America. The fact that the same mutation is found in all night-blind Briards indicates a founder effect, similar to how Canavan disease came to the United States from the Vilna Ghetto in Lithuania. The inbreeding that maintains purebred dogs perpetuated the visual problem, just as people partnering among themselves keeps a disease in a particular population group.

A Briard is a friendly animal, described by enthusiasts as "a heart wrapped in fur" and "a living fence." They excel at herding, guarding, tracking, and hunting, relying on intense senses of hearing and smell—perhaps to compensate for the night blindness that plagues some animals.

Eighth-century tapestries and scattered records from various times and places depict or describe Briards. On the battlefields of the French Revolution, the dogs used their natural talents to carry food, supplies, and munitions, and to sniff out the dead and dying. Historical rumor has it that Marquis de Lafayette gave Thomas Jefferson his first Briard, which he brought to the United States in 1789. The breed nearly vanished after World War I because the French Army depended on the dogs so heavily in combat zones.

The hardy Briards survived to become an official member of the American Kennel Club, in 1922. The breed's distinctive characteristics include a pair of dewclaws on each rear foot, making them halfway six-toed; a shaggy beard and eyebrows; and a tail with a small hook at the end, called a crochet. The dogs range in color from various shades of tawny to gray to black, and have two coats. The undercoat is thin and tight. The outer coat is coarse, hard, and dry, the hair lying in waves and, owners report, tending to shed readily and stick to everything.

Briards make wonderful pets. Adding to their historic skills on the farm and battlefield, the breed has starred on the small and large screens. Famous TV and film Briards include Tramp from *My Three Sons*, Rosie from the film *Dennis the Menace*, and Buck from *Married . . . with Children*. Half-Briards with notable theatrical credits include Fang from *Get Smart*, Stinky from *Dharma and Greg*, and Them from *The Addams Family*.

Back in 1988, the Briard breeder traveled several hours from central Sweden to Dr. Narfström's lab with all nine of her huge dogs crammed into a van. "I spent the whole day examining them and found that five had night blindness!" Narfström recalls, still excited. The breeder had noticed only the most severely affected pup, the one who kept banging into things in dim light. The pups' parents were sighted. The diagnosis: congenital stationary night blindness (CSNB), an older and broader descriptive term that includes Corey's disease.

Five out of nine night-blind pups may have been bad news for the distraught breeder, but it was the best possible news for a geneticist—or a research veterinarian with an interest in genetics. The litter was large enough for a trained eye to immediately see how the mutation was passed on. Whatever was wrong with the animals, it affected both sexes and could skip generations—important clues.

Narfström put the animals through a battery of standard vision tests: electroretinography (ERG) to measure retinal activity, exams with an ophthalmoscope, and maze testing. The breeder was so happy with Dr. Narfström's intense interest that she gave her a pup, starting the veterinarian's continuing love affair with the charming animals.

The next thing a geneticist does after spotting a trend is to confirm it in an expanded population. This is most expediently done with brother-sister matings, which sets up the inbreeding that more quickly reveals recessive inheritance. The sexually cooperative dogs didn't seem to mind being experimental subjects. Narfström bred the clumsiest dog to a visually impaired littermate, and they gave her eleven puppies. She knew Mendel's first law predicted that all of the offspring of two affected animals would likewise be night-blind—and they were. "I bred more litters and very quickly established that it was an autosomal recessive disease." This meant that the trait affects males and females (autosomal), and that carriers can have the mutation but not suffer night blindness because they also have a functioning copy of the gene (recessive). Blind Briards begat blind Briards because each dog had mutations in *both* copies of the responsible gene—those were the only offspring possible. But more tests were needed to discover exactly why the dogs couldn't see.

Next, Dr. Narfström crossed a few night-blind Briards with beagles, an easier breed to work with because they are small and have short hair and excellent eyes. Beagles have no visual conditions that might obscure the inheritance pattern of the Briards' disease. The trait would be easier to trace in the mixed breed. The resulting hybrids were smaller than Briards with shorter hair, but still shaggy. Each had one mutant copy of the gene, from the Briard line, and one normal copy, from the beagles. Then Narfström crossed the hybrids back to night-blind Briards, and saw the expected fifty-fifty ratio of normally sighted to night-blind offspring. She'd essentially repeated the famous experiments of Gregor Mendel, who crossed hybrid pea plants back to plants bearing the original trait to reveal autosomal recessive inheritance.

Narfström published the finding of congenital stationary night blindness in Briard dogs in the *British Journal of Ophthalmology*, writing, "It thus appears that the Briard dog may become a valuable model of human CSNB." The year was 1989, more than a decade before Corey Haas was born. The discovery of the blind sheepdogs was the first step toward his gene therapy.

Many nonhuman animals have featured prominently in the history of medicine. Mice are by far the most commonly used species, but less than a tenth of interventions that are promising in mice eventually find their way into human bodies. At some point, researchers usually switch to a larger animal model that better approximates the sizes of human body parts. Pigs stand in for human heart patients; horses are used to study arthritis; dogs to study muscular dystrophy. An animal model of intermediate size that has been important in the development of gene therapy for LCA is *Gallus gallus*—the chicken. A flock of Rhode Island Reds at the University of Florida serves as a model for Leber congenital amaurosis type 1 (LCA1), a more severe cousin to Corey's disease. The birds are completely blind by eight months of age.

LCA1 is like a cell phone that goes dark, stops working, and won't turn back on. In the chicken's eye, or the eye of a child with the same form of LCA, the rods and cones send the electrochemical message to the brain—*light! photons!*—but the signal doesn't get through, and the cells can't return to the resting state. Chicks with LCA1 are easy to spot. Wave a hand in front of them and they stand still. If they sense activity some other way, perhaps perceiving a vibration or smell, they might grow anxious, moving in circles, their eyes flitting back and forth. In contrast, their sighted barnyard mates focus on and peck at anything that might be edible.

A baby with LCA1 is harder to notice than a fluffy, disoriented chick. "There are only a few clues that an infant may have this disease," says Susan Semple-Rowland, PhD, principal investigator at the McKnight Brain Institute at the University of Florida in Gainesville. "Often parents will notice that their child doesn't seem to be smiling at or looking at faces. Children may also poke or rub their eyes, behaviors clinically known as oculodigital signs that may produce sensations of sight," she adds.

University of Florida researchers identified the cause of the chickens' blindness—a missing enzyme called guanylate cyclase, or GUCY1*B—at about the same time others found the same enzyme missing in children with LCA1. Like the Briards, the chickens were a fluke, the result of a spontaneous mutation in a gene that has a counterpart in people. "All of a sudden, the chicken became a valid model of a human disease," recalls Semple-Rowland.

She set up "rescue" experiments to prove her point. "The idea was that if we put the enzyme back in, with gene therapy, the bird's visual system should work. And it did. But it took ten years" of experiments. They used lentivirus (HIV) to deliver the *GUCY1*B* gene, which is too bulky to fit into AAV.

In chickens, the RPE is so tightly affixed to the retina that trying to slip in a gene between the two, as gene therapy would do with Corey, would wreck the entire multilayered structure. Fortunately, the chicken offered another route of entry—through the eggshell, or *in ovo*. Semple-Rowland, a cycling enthusiast with a fringe of reddish-brown bangs, short hair, and a robust laugh, talks about the approach in bursts of excitement. "We taught ourselves how to cut a tiny window into the eggshell, and microinject it at the stage when you can see the neural tube developing in the embryo. We could see the pretty blue dye go in. We injected a number of eggs with a lentivirus carrying the gene." The next step was to nurture the eggs. Like the cartoon elephant Horton hatching the egg of the lazy Mayzie bird in the Dr. Seuss book, Dr. Semple-Rowland and her coworkers faithfully illuminated, warmed, and rocked the chicken eggs injected with genes to cure their hereditary blindness.

On the eighteenth day, the researchers moved their subjects to a higher-humidity hatching incubator and awaited the big event. Fluffy chicks pecked their way out at the expected time. When the hatchlings were three to five days old, they were moved to the "animal housing unit," a dorm of sorts for poultry.

Then came the tests, some of which Semple-Rowland made up as she went along. First, the researchers placed each chick in the center of a drum with rotating striped walls. The contraption looked a little like a test to see if astronauts will throw up in space. Did the animals' eyes follow the revolving scenery? "I thought I saw signs that the chick was responding visually to the environment, but I didn't want to believe it. Scientists always doubt what they see—it's intrinsic to how we operate." She needed to test the bird's vision another way. "I did this simple little test, drawing little dots on a piece of paper. The chick, which was standing on the table, came over to the paper and started pecking at the dots. It was so exciting! You can put the gene in and get vision. Those were amazing chickens!"

They tested more chickens, in more ways. They added metal objects and brightly colored Skittle candies to the pecking surface. Treated chicks as young as three days old duly pecked at the objects, as did normal chicks. The untreated, blind chicks did not. Then, on day thirteen, the student running the tests discovered a uniquely avian behavior that the researchers hadn't considered. A bird called #2 began following the student when she moved. When she stood at the opposite side of the testing area, bent down and clapped her hands, the trusting #2 flapped and waddled straight to her, leaping off the platform into her waiting hands. The famed animal behaviorist Konrad Lorenz discovered imprinting when a group of adoring orphaned geese decided he was their mother; similarly, chicken #2 fell in love with the being it saw at a critical period in its development—the student. A blind bird could not have imprinted in this way.

The delightful videos of the imprinted baby birds make the experiments in the Florida chicken lab seem simple. However, they cap a quarter century of work and dispel anew the myth of the sudden scientific breakthrough. "It took quite a long time to build the vector, develop the injection procedure, and figure out how to hatch the eggs," Semple-Rowland says. She's looking ahead to clinical trials to treat infants who have LCA1, perhaps detected prenatally—just as she treated, and cured, the chicks.

One controversial advantage of using animals in experiments is that they can be sacrificed to reveal the extent of a correction, and possible adverse effects, before attempting the procedure on people. The Florida researchers sacrificed some chicks after six weeks. Dissection of their eyeballs revealed the minimum percentage of rods and cones—about 15 percent—that have to take up the healthy gene to restore vision for this form of LCA. Based on what the chickens told them, Semple-Rowland thinks that the best approach for babies with LCA1 may be to treat several parts of the eye, minimally. Corey's results have perhaps been so spectacular because LCA2 affects the RPE cells, and not the rods and cones directly, as in LCA1.

Because one RPE cell nourishes many rods and cones, a limited correction goes a long way. "LCA2 is the easy one. The AAV dudes did that one really fast. It will open the door to allow future gene therapies to move faster into the clinic," says Semple-Rowland.

Unfortunately, the chickens enjoyed their vision for only about four months, and then they began to stumble. A chicken lives three to four years, so the human equivalent of their four-month restoration of sight—about seven to eight years—wasn't good enough.

"I can't imagine giving somebody sight and then saying, 'In a year or two you won't be able to see again,'" says Semple-Rowland. That's the 800-pound gorilla in the room when it comes to the successful gene therapy for Corey and the others. *Is it really a forever fix?* So far, the animal studies suggest that it is.

17

LANCELOT

WITH ITS MANY DISTINCTIVE BREEDS, THE DOMESticated dog is a geneticist's delight. Our centuries-long interference in their reproduction has selected and shaped the dog genome into myriad variants so ill-adapted that they likely would never have happened by natural selection. In many breeds, coveted traits came along with health problems that have persisted as continued inbreeding, often from only a few animals, created a human-driven founder effect.

Just a decade ago, it took years to zero in on the gene behind a dog disease. A researcher would narrow down the part of a chromosome that the members of a breed with a particular condition uniquely share—such as breast cancer in Springer spaniels or an enlarged heart in Newfoundlands. The next step was to identify the part of the chromosome that is abnormal among members of a *different* breed that has the same condition, but less frequently. Little by little, as the researchers probed different breeds for the same disease, a small part of the dog genome emerged that all affected animals shared. Eventually, they'd discover in that chromosomal neighborhood a gene identified by the canine genome project whose function, if compromised, could cause the symptoms. Sequencing of the dog genome—accomplished first in a female boxer named Tasha, in 2004—has greatly accelerated gene discovery in dogs.

We share a variety of medical conditions with our canine friends. To study retinitis pigmentosa, Gus Aguirre and his colleagues at the

University of Pennsylvania use English mastiffs and Irish setters. Mini schnauzers, Siberian huskies, and Welsh corgis also get versions of the disease. Alsatian dogs and German shepherds inherit giant axonal neuropathy (GAN), starting to stumble at about fifteen months of age as their limb muscles weaken, just as Hannah Sames started to do at about the same age. At Stanford University, a ten-year experimental journey with Doberman pinschers and Labrador retrievers led to identifying new drug targets for human sleep disorders. (The work has taken so long because of the challenge of breeding narcoleptic dogs. Sexual stimulation so excites them that they fall fast asleep before they can complete the act.)

Dogs are especially good models of psychiatric tendencies and disorders. About 40 percent of breeds have a behavioral quirk. Some have "canine compulsive disorder," which is like human obsessive-compulsive disorder, the "CCD" instead of "OCD" presumably because dogs don't obsess as we do. CCD takes the form of tail-chasing in bull terriers, and licking the legs until the fur falls off in golden retrievers, German shepherds, and great Danes. The intense licking is akin to repetitive hand-washing in humans. English Springer spaniels and cocker spaniels are prone to fits of rage that might reflect a canine version of bipolar disorder, while Labrador retrievers have attention deficit disorder. Alas, we have likely caused some of these problems. Nearly half of border collies, for example, have "noise anxiety," nearly jumping out of their skins at a clap of thunder or a slammed door. Breeders selected the extreme sensitivity to sound that enabled the animals to hear and respond to commands from far downfield, so that they could herd. Their descendants, no longer in the field but in the living room, are skittish at any sudden noise.

Animals used in preclinical trials have a very specific role: to test an intervention. Researchers test whether a drug is toxic, or a surgical procedure too damaging, in a zebrafish or mouse before trying it on children. But even a terrific animal model such as the dog can't always predict what will happen when the treatment is translated to people. This was the case for hemophilia B, the rare form of the "bleeder's" disease that affected the royal families of Europe, starting with a mutation in Queen Victoria.

Katherine High, MD, director of the Center for Cellular and Molecular Therapeutics at the Children's Hospital of Philadelphia, has spent most of her career doggedly pursuing gene therapy for hemophilia B, which would free patients from frequent, costly infusions of the missing clotting factor IX. Kathy High is also part of Corey's medical team, and has done everything from designing viral vectors to designing clinical trials. Minutes before her talk at the American Society for Gene and Cell Therapy annual meeting in Washington, D.C., in May 2010, she stopped on her way to the podium at the front row, where Corey sat with his parents, awaiting Jean Bennett's presentation when he would take the stage. After greeting Corey, High leaned over to whisper to his mom.

"Nancy, I have to tell you something," she said mysteriously, as Nancy looked first puzzled and then a bit worried.

"My talk is rated PG13. I have to mention semen." She looked at Corey. "Will that be okay?"

Nancy nodded yes and laughed, patting her son's head. Corey was clearly more concerned with collecting ribbons to attach to his name badge than with listening to a science talk, and by the time the "semen" reference came around, he was fighting to stay awake. But the rest of the crowd was riveted.

Kathy High took listeners through an account of hemophilia B gene therapy called "The Clot Thickens." After many experiments with Cairn terrier–beagle mixes, it was on to humans. But in late 2001, an intriguing finding emerged in a clinical trial at CHOP for hemophilia B. In one man, the virus (AAV) that carried the clotting factor gene into his liver took a detour to his semen. If the vector wound up inside sperm cells, then its cargo could, theoretically, be ejaculated into the next generation. Such "germline gene therapy" was a big enough ethical no-no for the FDA to instantly halt the trial, especially since it was so soon after Jesse Gelsinger's death, also delivering gene therapy to the liver.

Dr. High was upset but intrigued. An appearance of the therapeutic gene in semen hadn't occurred in the experiments with the dogs, when the researchers had meticulously tested many tissues to see where the vector went. "It was a complete surprise. We fractionated the semen and the vector was not in the motile sperm, but in the seminal fluid," she told the crowd. That meant the gene wouldn't be passed to the next generation after all, but High was still concerned. Jesse Gelsinger's case had shown that knowing where a gene therapy vector might go, other than its target, is vitally important. How did the vector carrying the clotting factor end up in the seminal fluid?

At this point Dr. High looked at the audience with a mischievous smile. She seems a proper professional, not at all the type to giggle over the sexual or scatological. Then she let the crowd in on the joke. "I had bad karma from skipping lectures in medical school. I was getting my wisdom teeth removed, and since I was sure I wasn't going to be a urologist, I scheduled the extraction for when we'd be covering the male reproductive tract. Now I had to learn it. Fast."

While High was trying to figure it all out, more men in the trial started to show viral vectors in their semen. The researchers closely scrutinized the data: the vector and its cargo left the semen gradually, and the higher the dose, the longer it took. Then Dr. High noticed something strange. "We found that the younger the man, the faster the vector cleared." She stopped talking for a few moments as the audience mulled this over, and, little by little, nervous smiles broke out. After a beat, High continued, "So that led to another hypothesis, which we checked in rabbits. The kinetics of clearing depends on the frequency of ejaculation."

The escape of an occasional virus into the reproductive tract did not derail the clinical trial for gene therapy for hemophilia B, which will continue for another fifteen years. But the errant vectors pointed out that even the best animal models don't always completely replicate the human condition.

It took several years to figure out why the Briard sheepdogs that the breeder brought to Kristina Narfström couldn't see. She assigned two of her graduate students at Linkoping University, Anders Wrigstad and Sven Erik Nilsson, to investigate the cause of the blindness, while she continued studying optical and brain problems in Abyssinian cats, Persian cats, Tibetan terriers, Polish sheepdogs, and wirehaired dachshunds.

The Narfström-Wrigstad-Nilsson trio published a series of papers on their anatomical findings on the Briards, taking turns as first author. A pair of papers published in *Experimental Eye Research* in 1992 described the abnormal structure and function of the eyes in what the team still was calling congenital stationary night blindness. The researchers flashed lights at five seven-to-twelve-month-old Briards and looked at the electrical activity in the layers of the retina. As was true with Corey's retina, activity was nil.

When the researchers dissected the retinas of four very severely affected pups born to a brother and sister from the original litter, they found a number of abnormalities. The RPE looked ragged, pocked with dark spots amid fatty droplets. The investigators didn't realize it at the time, but these bubbles housed the form of vitamin A (trans retinyl esters) that must be converted to 11-cis retinol for the rods to regenerate the visual pigment rhodopsin. The fatty retinyl esters built up because the eyes lacked the enzyme to convert them to the usable cis form. (*Trans* and *cis* refer to the positions of atoms with respect to each other in a molecule. Atoms in cis are on the same side of the axis formed by a molecule's backbone structure, whereas atoms in trans lie on opposite sides. Bicycle pedals provide an analogy: one up, one down, is trans. If both pedals could be up or down at the same time, that would be a cis configuration. Identical molecules, one cis and one trans, may react differently.)

The photoreceptors that lie nearly up against the RPE also were abnormal in the dissected eyes of the blind dogs. The cones looked okay, but the toothbrushlike ends of the rods were shaggy and disheveled, not their normal neat array. But, importantly, the rods and cones were *there*. So whatever was blinding the Briards wasn't congenital stationary night blindness after all—not if the eyes still had rods and cones. This was good news. Perhaps the researchers could correct the blindness if they could resurrect the support system in the RPE.

At the same time that the Swedish researchers were dissecting and describing the eyes of the blind Briards, researchers at the Laboratory of Retinal Cell and Molecular Biology at the National Eye Institute in the United States were proceeding in parallel, looking at the responsible genes and proteins. R. Michael Redmond and coworkers identified and named the RPE65 protein in 1993. Although it didn't match anything in the protein databases at that time, they subsequently found it in a variety of mammals, birds, and frogs. Since evolution tends to keep whatever works, the mystery protein must be important. Looking at eyes of other species, Redmond's group discovered that RPE65 protein is tightly regulated throughout development and dots the labyrinthine membranes inside normal cells.

The next chapter in the unfolding Briard story appeared in a journal called *Documenta Ophthalmologica* in 1994. Wrigstad, Narfström, and Nilsson examined seven eyeballs from two generations of dogs taken between five weeks and seven years. A time course emerged. In five-week-old eyes, only the rod outer segments, the toothbrush bristles, were disordered, but by three and a half months, dark spots riddled the RPE. By seven months, the rods were actively degenerating, mostly at the edges of the retina. Finally, by seven years, most of the rods at the periphery of the retina had vanished, and were sparse elsewhere. The degeneration had spread to the inner layers of the retina. The researchers renamed the condition hereditary retinal dystrophy, denoting a slow wasting process rather than an absence of cells from the start or their sudden destruction. It was an important distinction for developing a treatment.

Armed with this vivid description of what was happening in the Briards' eyes, the researchers next had to look for the molecular underpinnings. That would require collaboration with molecular geneticists. So later that year, 1994, Anders Wrigstad contacted Andreas Gal, PhD, a researcher at the Institute for Human Genetics at the University Krankenhaus Eppendorf in Hamburg. Dr. Gal and his graduate student Andres Veske would search for the blindness mutation in the Briard blood samples that Narfström sent them. Over the next few years, the team of Swedes and Germans ruled out a few candidate genes. About that time, Wrigstad mentioned Leber congenital amaurosis as a possible diagnosis. He was the first to link the Briards' disease to Corey's, Narfström says.

In 1997 and 1998, a strange convergence of research pathways nailed Wrigstad's LCA hypothesis.

First, the Narfström-Gal team finally discovered the mutation: four DNA bases in a row missing in the gene that encodes Redmond's RPE65 protein. At about the same time, Gal and other coworkers linked mutations in the *RPE65* gene to severe retinal disease in children, publishing the work in the prestigious journal *Nature Genetics*, complete with a "News and Views" commentary. Then a report in the equally prestigious *Proceedings of the National Academy of Sciences* implicated *RPE65* mutations specifically in LCA in children. This was the paper Anne Fulton coauthored, which figured in her diagnosis of Corey's visual deficits.

All of these developments were huge news, because the Briards now provided a large-animal model to test a gene therapy. It didn't even require any genetic manipulations—Mother Nature made the dogs blind. But in hindsight, Dr. Gal might have been better off reining in his enthusiasm.

"We were trying to publish our results. Meanwhile, Andreas Gal presented the results at the annual meeting of the American College of Veterinary Ophthalmologists, letting the cat out of the bag. He stood up and told everyone the defect, too early," says Narfström, her voice stumbling over the memory of the event that altered her career. For in the audience sat a research group from Cornell University that had also been working on the lipid buildup in the Briards' eyes and were close to finding the mutation themselves, in dogs from the United States. Back in 1975, Narfström had spent six months in the lab of their leader, Gustavo Aguirre. More recently, she'd sent them blood from four of her blind dogs, so they could compare the Swedish and U.S. mutations to be sure they were working on the same disease.

Both research groups were coming to the same conclusion, but in science, the first publication marks the turf—not a mention at a meeting. When a race is so close, the peculiarities of journal publication schedules can obscure who crossed the finish line first. "Several labs were diligently looking for the causative gene. Gal's group beat everyone by presenting this information in public. As we were quite far along in also doing this work, we went ahead and published the work, and 'beat' him by one year in the actual date of the publication, 1998 versus 1999," explains Aguirre. That paper appeared in *Molecular Vision* and includes Narfström as a coauthor because she supplied the dog blood, and credits Gal and associates with identifying the mutation at the veterinary meeting, at the start of the article.

In between the October 1998 Aguirre paper and the April 1999 one from Gal and Narfström came a perhaps even more important report, from Michael Redmond's group at the National Eye Institute. Using mice, they showed that RPE65 was indeed the long-sought enzyme that turns the useless trans form of vitamin A into the cis form that the rods need.

The pace of the research accelerated. In 2000, Aguirre and two coworkers, Gregory Acland and Kunal Ray, filed a patent for a carrier test to enable breeders to avoid blind pups in their carefully conceived litters. That same year, Gal's group described the disease in young children in an ophthalmology journal, calling it a specific form of LCA: "*RPE65* mutations should be suspected in infants who appear to be blind in dim surroundings but react to objects in bright illumination and have nonrecordable rod ERGs and residual cone ERGs." It was a perfect description of Corey Haas.

As soon as the researchers unveiled *RPE65* as the cause of the Briard's blindness, planning began for gene therapy for LCA2. It would be attempted first in dogs, in research groups on opposite sides of the Atlantic. LCA2 was a perfect candidate for gene therapy. In affected retinas, particularly in the young, not all of the rods unravel, and those that do, do so slowly. Saving the rods would be like restoring an erased iPod—the basic structures are still there, just unable to function. Another reason to try gene therapy on LCA is that the eye is naturally off-limits to the immune system, which is why corneal transplants need not match between donor and recipient. A dangerous reaction like the one that killed Jesse Gelsinger was therefore unlikely, if not impossible. Plus, it's much easier to see results in an eye than in a liver or brain.

For Kristina Narfström, developing *RPE65* gene therapy for dogs was, to quote the baseball great Yogi Berra, déjà vu all over again. She was afraid she'd be scooped on the gene therapy, just as she felt she had been on the discovery of the *RPE65* mutation that causes LCA2, even though she'd been a coauthor on another group's report.

The team that accomplished the gene therapy for Corey's disease first, in "a canine model of childhood blindness," was a Who's Who that included Cornell's Acland and Aguirre; the "AAV guru" Bill Hauswirth, PhD, and Sam Jacobson, MD, PhD, from the University of Florida, who would soon lead the first human gene therapy trial for the disease; and Jean Bennett and Al Maguire from Penn, who would treat Corey and the others in his clinical trial. The paper appeared in the May 2001 issue of *Nature Genetics*, reporting on experiments begun the previous summer, and the author list did *not* include Kristina Narfström.

On July 25, 2000, three dogs received the gene therapy. The first patient was Lancelot. He had been a very sad puppy, recalls Jean

Bennett. "He'd bump into doors, and he couldn't find his water bowl. He became very timid and just sat there."

For the gene therapy procedure, each dog was anesthetized and cushions used to gently position the animal on the operating table. Drapes exposed the right eye. Then the surgeon introduced instruments into the eye in two places through the sclera, the white part. One hole admitted a lit probe to view the procedure through an operating microscope. Through the second hole the surgeon inserted a custom-made glass micropipette, or "microinjector," which was a tiny tube about the width of a hair. The tip of the micropipette gently perforated the retina and slid underneath it, reaching the "subretinal space" where the RPE cell layer abutted the tightly packed photoreceptors. The microsurgical invasion locally detached the retina, forming a little bulge called a bleb that healed within a day, and released droplets of gene-loaded viral vector, using precise dilution factors and volumes known from previous experiments. It all took just a few minutes.

The plan was to assess the dogs in about ten days on a mobility course set up in the hospital hallway. The number of times each dog avoided each pylon or box placed in its path generated quantifiable data, something vital to show success for the next grant submission. But it turned out that the dogs showed the gene therapy worked without any tests at all.

"The dogs are spinning! They're spinning!" shouted the lab tech to Dr. Aguirre after checking the animals a few days after the procedure. With the world suddenly appearing in each dog's right eye, spinning was an instinctive way to expand the visual field to see as much as possible. The dogs looked happy, the tech said.

Jean Bennett smiles with a twinkle in her eye as she remembers the excitement in the aftermath of the gene therapy. "At two months, Lancelot could see. When we saw the dramatic results, we knew we wanted to do it on humans. And Corey was born right about when we first noticed that Lancelot was getting better."

The spinning dogs excelled on the assessments written into the protocol. At four months, ERG testing showed electrical activity at

16 percent normal level, with the pupils constricting in bright light about half as much as normal. Before the treatment, response had been nil. It was also clear from the "qualitative visual assessment"—simply watching the experimental subjects lope down the object-strewn hallway chasing Dr. Jean, her wavy mane of red hair and white lab coat tails flapping—that the dogs could see well enough to navigate. They avoided obstacles in front and to the right, but not to the left, while the untreated control canines still crashed into everything. In her presentations at professional meetings, Dr. Bennett shows a slide of a bespectacled Lancelot holding open the May 2001 issue of *Nature Genetics* that reported on the successful-beyond-anyone's-expectations gene therapy for LCA2.

Nature Genetics, the University of Pennsylvania, and Cornell University issued news releases to publicize this fabulous example of the compassionate use of animals in research, as Lancelot made public appearances. It is kosher for a reporter or blogger to borrow verbatim from a news release, and that may be how Kristina Narfström's contribution was, for a time, left out of the story that led to Corey.

"This is a perfect example of how research in animals helps both the animal—in this case the dog—and the human patient," Dr. Aguirre was quoted in Cornell's version of events. But the news release went on to say, "whose laboratory discovered the *RPE65* gene defect in briard dogs in 1998." According to the published record, that was true.

Sometimes researchers check a news release about their work for accuracy before it goes out to reporters everywhere, but often a scientist is too busy to pay much attention to details of wording. Reporters on a deadline, especially those who don't regularly cover science, might not have the time to check details in the technical paper behind the news release. Errors of omission are the hardest to spot. Without reading the second paragraph of the 1998 *Molecular Vision* paper, a reporter couldn't know that researchers in Sweden had actually identified the critical mutation first. The misattribution continued. An article from the National Eye Institute published a few years later also credits the Aguirre team with the discovery, doesn't mention Narfström's group, and even quotes Redmond calling discovery of the *RPE65* mutation in Briards a "no-brainer."

At about the time of the May 2001 *Nature Genetics* publication, Lancelot appeared on Capitol Hill. He charmed congressional representatives at a luncheon, shaking paws and happily retrieving tidbits tossed near his good side. In the background, a video showed him stumbling around sightless, pretherapy, so his antics made a powerful visual aid for the appeal to increase spending for gene therapy. Aguirre pointed out to the crowd how Lancelot stood with his right side, his seeing side, to the crowd, and how he paid attention to the speakers at the podium. If he'd brought in a mouse treated with gene therapy, Aguirre joked, most people wouldn't be eager to pet and adopt it, as they did the friendly dog. But mostly the researchers were relieved that Lancelot—who hadn't been raised as a pet—didn't pee on the beautiful Oriental carpet.

As convincing as Lancelot's performance before Congress was, and although researchers were already screening human patients for the gene therapy, his dramatically improved vision was just the beginning of the animal-research part of the story. More dogs needed to be treated and followed to work out the details of how best to deliver the gene. How many injections at once? Where to deliver them? What is the minimal effective dose? How long does the improvement last?

Narfström's group published a series of papers on dogs they'd treated with gene therapy in 2001 and 2002, but without benefit of the PR machine of a major journal or university. Their first paper appeared in *The Journal of Heredity*—a nice-enough publication, but one devoted to "organismal genetics." A report in *Nature Genetics* might inspire coverage in *The New York Times* or *The Washington Post.* But *The Journal of Heredity*'s covers in that year, 2003, featured pigs, a spider, a cactus, a horse, a kangaroo, a tomato, a dinoflagellate, and a whale—not exactly exciting mainstream media images. Still, the journal allowed Narfström the space to describe what her group actually did to the dogs, rather than referring to other reports for the details, like the *Nature Genetics* paper did.

Narfström and her students followed five treated dogs of various ages: Candy, Bonus, Perdita, Millie, and Rex. They monitored the dogs for eleven months, extending the four-month analysis of the first gene therapy trial. (Bonus would accompany Narfström to the University of Missouri in 2002, where he would happily contribute to the Briard gene pool in the United States.) Each dog had received 70 to 100 microliters of fluid containing the AAV vector carrying its *RPE65* gene in the right eye, and a jellyfish's green fluorescent protein gene, also delivered aboard AAV, to the left eye, as a control to see if the vector was actually going where it was supposed to. But Millie's left pupil mysteriously constricted during the procedure and couldn't be injected—which turned out to be fortuitous.

The results were every bit as dramatic as those for the original cohort of spinning pups. "Four weeks after surgery there was a marked change in the visual behavior of the previously blind dogs," the researchers wrote. The dogs were more alert, didn't need sound or smell to navigate, and the flicking back and forth of their eyes (nystagmus) stopped by ten weeks. Not only did the apparently safe treatment work in dogs of all ages, but it seemed to help the control eye too—even in Millie. She hadn't received the sham injection in her left eye, so it couldn't be the intervention itself that improved vision.

This unanticipated response in the second, untreated eye was what Ethan Haas saw on a sparkling October day shortly after Corey's gene therapy. A month after the surgery, back home in the Adirondacks, Ethan and Corey were in the backyard, raking leaves and brush into piles near the woods before the inevitable six months of snow descended. Corey looked up from the task, a bit winded yet exhilarated with the scent of the leaf litter. Ethan had always called him "owl eyes," because his pupils would grow so wide outdoors, as if trying to soak up all the light rays they could.

But now something was different. For the first time ever, Ethan saw the full blue of his son's eyes, as the pupils constricted in response to the tawny autumn light. Ethan noticed that the boy's eyes were a startling blue—both eyes, even though he'd been treated in only one.

When Narfström noted improvements in the untreated eyes of her dogs, she thought back to a 1999 paper from the Cornell-Penn group. They'd shown that after sending the gene for green fluorescent protein into the retinas of healthy dogs and mice aboard AAV, the telltale green showed up not only in the retinas, but also in the optic nerves that lead to the brain, as well as *in* the brain. This finding raised an intriguing possibility: could the eye be a route to treat a brain disease, like Parkinson's or Alzheimer's? More immediately, Narfström knew that what she was seeing in her recovering dogs was that the RPE protein—not the gene, but the protein—was spreading beyond the treated eye to its untreated mate.

Restored vision, even just a little, in the second eye was an unexpected gift. But the improved vision peaked at three months. Narfström went back to perfecting her technique.

Her next results appeared in *Investigative Ophthalmology and Visual Science* in 2003, and then *Documenta Ophthalmologica* in 2005, also journals far from the path of the average beat reporter. But this time she teamed up with the group from the National Eye Institute investigating *RPE65* eye disease in people. They performed gene therapy on eleven blind beagle-Briards and compared them with three control dogs. Results were more encouraging. Again the improved vision peaked at three months, but this time it ebbed more slowly from the untreated eye, lasting for nine months. These papers also examined fourteen eyes apart from their owners.

Sectioning eyeballs sounds grisly, but it reveals what is going on in a way that couldn't happen in a human clinical trial. Dissections showed that the gene therapy was reaching its target—the RPE—and that near the injection site, the little fatty bubbles had vanished. This didn't happen in the control eyes that had received the vector with the dummy glowing green cargo. The gene therapy was working, and the researchers could now see how, at the cellular and molecular levels. What's more, despite some minor inflammation in the middle layer of the retina and damage at the injection site, the eyes healed and the dogs, some now three years posttreatment, could really see.

In 2005 came a milestone paper from the Penn-Cornell group and others. The title said it all: "Identifying Photoreceptors in Blind Eyes Caused by *RPE65* Mutations: Prerequisite for Human Gene Therapy Success." Basically, kids like Corey needed to have at least some rods and cones for the repaired RPE to restore vision—as was the case in dogs and mice. This paper appeared in the ultraprestigious *Proceedings of the National Academy of Sciences.* The researchers scrutinized the eyes of eleven people with LCA2 using optical coherence tomography, not one of Corey's favorite tests. OCT reveals the retinal layers in glorious colors, even the topography of the rods and cones. It's a good stand-in for removing and slicing up the eyeball.

As in dogs and mice, the eyes of the patients still had some photoreceptors, although vast areas in the central retina were depleted of cones as rods ebbed away toward the periphery. Not surprisingly, the older the patients, the blanker their OCT maps. Again, like a dead iPod, all the rods and cones needed was to be recharged, and gene therapy could do that. What's more, OCT might enable surgeons to tailor *where* they introduce genes, to exactly where the photoreceptor layer appears thickest in a particular patient. The paper called gene therapy in people the "logical next step."

In fact, planning for clinical trials was already well under way, at more than one institution. Meanwhile, the Cornell-Florida-Penn team continued treating and assessing dogs, varying parameters and cataloging effects. They tried slightly different viruses, compared human and canine *RPE65* genes, and stitched different control regions into the DNA that affected which tissues use the gene. They measured degree of correction and duration of response. Their next set of experiments, published at the end of that year, 2005, restored vision to twenty-three of twenty-six treated dogs. One eye in a dog that didn't see held a subtle but telling clue—it showed the biochemical correction, but the dog's vision hadn't improved. Was a threshold level of correction required to see better? Did the number of injection sites make a difference? It was important to know how many gene injections a person might need, and where they should be placed, for the therapy to exert a noticeable and sustainable effect.

In 2007, the Cornell-Florida-Penn team tackled yet another compelling question: How does the brain respond to suddenly being able to sense light? In dogs, fMRI scans showed that the visual cortex can process new visual information into meaningful images. In people with Corey's disease, extrabright light stimulates their brains more than dim light, which explains the boy's penchant for staring into lit bulbs as a baby. In LCA2, therefore, in dogs and people, enough of the visual equipment is still present and can be coaxed to respond. This work appeared in yet another prominent journal, *PLoS Medicine.*

In 2008, Narfström contributed other vital observations and insights, but in a book chapter. She noted damage if the injection was directly under the macula, and discovered that the photoreceptors continue to degrade in areas away from the injection. One dog, four years after surgery, had completely lost the photoreceptors at the edges of the retina. "This indicates that although there is a long period during which gene transfer can be effective in rescuing retinal function, eventually retinal degenerative changes would make such therapeutic intervention fruitless," she wrote.

Would people need "booster" gene therapies?

18

SUCCESS!

ANIMAL EXPERIMENTS DON'T END WHEN HUMAN clinical trials begin—preclinical and clinical research on the same disease often overlaps. As research to perfect the eye gene therapy on dogs continued, the first young adults received it, in late 2007 and early 2008, in London. When Ethan and Nancy first learned of the gene therapy for Corey's disease from Dr. Fulton in July 2006, she mentioned that the first clinical trial was being planned, in the UK. "We heard about the London work. We thought about getting our passports and going there, but we figured the research would take years," Ethan recalls. When an excited Dr. Fulton mentioned gene therapy again at the end of June 2007, Ethan and Nancy paid more attention. This was a different clinical trial, she said, and the gene transfer would take place in Philadelphia—a five-hour drive compared with an ocean away.

It all seemed too good to be true. Like most people, Nancy and Ethan had never heard of gene therapy, and had to suddenly process so much information that they procrastinated following up until early 2008. By the time they finally called Kathleen Marshall, the clinical research coordinator at Children's Hospital of Philadelphia, the first three patients, all young adults, had already received the gene therapy there. Each patient had the worse eye treated, the first patient in October 2007 and the most recent just a few weeks earlier. The researchers already knew it had worked—all three patients had improved vision within two weeks, and their back-and-forth eye movements had stopped, too.

A few weeks later, the Haas family visited CHOP, where Corey had preliminary vision tests and met the research team. In April, Jean Bennett talked to the media about how well the first three patients were doing. "We expect improvements to be more pronounced if treatment occurs in childhood, before the disease progresses," she told them. While the researchers were busy planning the inclusion of children in the clinical trial, the Haases heard nothing further until the summer of 2008, when the phone call came.

"Hi, Ethan, it's Kathy Marshall. I've got a quick question. When will Corey turn eight?"

"September twenty-second," answered Ethan. "Why?"

"We have a spot open for September twenty-fifth. Do you want it?"

~

It is common in science for more than one research group to tackle the same problem. LCA2 was such an obvious candidate for gene therapy that several teams attempted it at about the same time. But the multiple gene therapy trials for LCA2 weren't really redundant. They were necessary to provide the numbers that will help to establish the procedure as a standard treatment. They also tested variations on the gene therapy theme. How many billions of viruses to deliver? Is it better to treat in the area of the fovea, where photoreceptors are densest, or not? How much fluid can an eyeball hold? How young can patients be treated? How best to assess success? With more trials, the research teams refined their approaches.

The first clinical trials of gene therapy for LCA2 came from three groups, each treating three adult patients. Seven of the nine patients enjoyed improved vision very soon after the procedure. The timing and institutions overlapped. Group 1 came from CHOP and the University of Pennsylvania; group 2 from University College London/Moorfields Eye Hospital/Targeted Genetics; group 3 from the University of Florida and the University of Pennsylvania.

When the groups published their results—groups 1 and 2 in the May 22, 2008, issue of *The New England Journal of Medicine* and group 3 in October in *Human Gene Therapy*—the PR departments at the various institutions quickly took credit. The UK contingent claimed a "world first" while the CHOP-Penn camp touted the "first gene therapy for a nonlethal pediatric condition." "I think we redeemed gene therapy, in a way," Corey's surgeon, Al Maguire, told *US News and World Report*.

No matter who did what, when, and to whom, the results were astonishing. "In each clinical trial, there has been spectacular restoration of vision," said Bennett at the time. At the helm in London were geneticist Robin Ali, eye surgeon James Bainbridge, and retinal specialist Tony Moore. The London group's three patients were between ages seventeen and twenty-three. One of them, eighteen-year-old Steven Howarth, appeared on the BBC demonstrating his newfound ability to see the frets on his guitar. "I used to be shy and silent. Now I have confidence to walk around on my own," he said. Added Dr. Bainbridge, "It's tremendously exciting to see that this technique is safe in an extremely fragile tissue and can improve vision in a condition previously considered wholly untreatable."

In the competitive world of science, publishing first is often a goal, but sometimes a wait is worthwhile, because the researchers learn more. This was the case for the LCA2 gene therapy group led by William Hauswirth, PhD, from the University of Florida's Powell Gene Therapy Center, and Samuel Jacobson, MD, PhD, from the Scheie Eye Institute at the University of Pennsylvania. The trial took place in Gainesville, Florida, and a woman and two men in their early twenties had the gene therapy. Their stories are as compelling as Corey's.

Dale Turner, a resident of Toronto, was twenty-two when the Hauswirth-Jacobson team treated him in 2008. As a boy, he'd lived in a world of pale, blurry images, trapped within a narrow visual field. When he was finally diagnosed at age eight, his doctors predicted he'd be totally blind within two years. Although the gene therapy on his completely blind right eye reached only 4 percent of his retina, the viruses must have hit their mark, because Dale's response was spectacular. "Three days after the surgery, I got to go outside and I took off my sunglasses and peered into the sky. I had seen the sky before, but never like this, never with vibrant color," he said. For the first time, he could take walks with his girlfriend at night without crashing into things. He currently attends law school. When Dale returned periodically to Florida for checkups, a technique called high density microperimetry hinted at an unexpected response—newly dense retinal regions. Was his treated retina actually rejuvenating?

It was another patient in Dale's clinical trial who had perhaps the most amazing response to date, whose improvement confirmed what the researchers suspected was happening to Dale, although it took months for anyone to realize it. Her response also taught the researchers a powerful lesson in choosing ways to assess outcomes of experiments, as had the spinning dogs: not all the data come from blood tests and brain scans, eye charts and obstacle courses. "Patients tell you a lot—you have to listen to what they say," says Hauswirth.

What the patient reported was deceivingly simple: "A month ago, I began to read the green numbers on our car clock for the first time. I saw the numbers better when I looked above the clock," the puzzled young woman had reported. Hauswirth and his colleagues were even more ecstatic than their patient. They realized what her ability to read above center meant—that her treated eye was actually developing a brand new site of greatest visual acuity, a second fovea. Hauswirth compared her pretreatment retina to a map of the Pacific showing only a few scattered islands—regions where rods and cones persisted. Now she was growing new islands of photoreceptors. What began as a theoretical hypothesis—that gene transfer could replace RPE65 protein and restore vision—was now clinical reality.

Growing a second fovea was the best thing that could possibly happen. The fovea is a tiny pit at the back of the eye, an indentation in the retina no more than a millimeter across. Yet its importance is out of proportion to its size, for several anatomical reasons. The cones at the very center of the fovea are different than elsewhere, sleek and sculpted in a way that enables them to pack very tightly. Here, each photoreceptor connects to its own ganglion cell, the neuron that leads to the optic nerve, so that even though the fovea accounts for less than 1 percent of the retina, it delivers half of the visual input to the brain. Elsewhere in the retina several photoreceptors may share a single ganglion-cell conduit to the optic nerve. The fovea is a pit because it is stripped down to the barest essentials needed to signal the brain that light is coming—it lacks blood vessels and some of the intervening cell layers that cushion the rods and cones in the peripheral regions of the retina. Moving the eyes when reading enables the tiny fovea to capture the images of the letters. Without the eye moving, the fovea covers only about an inch of type at a time.

For the young woman, seeing the green numbers on her car's clock was just the first sign that the gene therapy had worked. Her new visual world came into focus gradually but surely, as her brain caught up to what had been done to her eye. "We don't fully have our heads around this. It took nine months for her brain to realize she had a useful region of her retina that was off-center, but better than the central fovea. Her brain rewired over nine to twelve months and decided that if light was dim, the eye would shift. Her brain decides from the image which part of the retina she uses. She doesn't know she is doing this, but her brain does. The two other patients are giving indications that this is happening to them, too," Dr. Hauswirth told a crowd at the American Society of Gene and Cell Therapy annual meeting in May 2010.

Other research groups have joined the effort to perfect gene therapy for LCA2. Just two weeks before Corey's initial gene therapy, Shalesh Kaushal, MD, PhD, at the University of Massachusetts in Worcester, J. Timothy Stout, MD, PhD, MBA, and Peter Francis,

MD, PhD, at the Casey Eye Institute at Oregon Health and Science University and Applied Genetic Technologies Corp., started a phase 1/2 safety and efficacy trial. In Israel the Hadassah Medical Organization launched a trial with ten patients in early 2010. And the FDA application to treat Corey and others' second eyes was filed on September 22, 2010, his tenth birthday and the first day of his two-year follow-up exams at CHOP. The first three young adults in the CHOP-Penn trial each had their second eye treated in the spring of 2011, and are doing well.

As time goes on it appears that the LCA2 gene therapy is indeed sustained, for an impressive percentage of the patients. Years out they're still seeing better, with improved light sensitivity, wider visual fields, and perhaps most telling, the ability to do normal, everyday things, like see a lit dial on a dashboard or a firefly's glowing rear. "Our best responding patient can see sixty-thousand-fold dimmer objects than before treatment. They've gained back all they could have, but no one has reached one hundred percent recovery, because they didn't start out with all the rods and cones," explains Hauswirth. And that's why the younger a person is treated, the better—there's more to save.

Getting a clinical trial for a gene therapy up and running is enormously more complicated than doing so for a conventional drug, especially since no such therapy had ever been approved for marketing. "Gene transfer is riskier and there's a lot more to consider. In addition to the things considered for drugs, you have to determine which vector to use, how much, where it goes, if you can direct where it goes, the time course, the level of gene expression, what happens if it goes somewhere not intended, and toxicity. With aspirin, a side effect could be an increase in liver enzymes. With gene therapy it could be worse," explains Jennifer Wellman, who led the herculean task of guiding Corey's clinical trial through all the regulatory and oversight committees. Wellman, with her long, straight brown hair and sparkling energy, looks like a senior in high school.

She has a master's degree in microbiology, and has learned enough on the job to have the equivalent of a PhD. After five years managing the vector core facility at Avigen, a biotech company in Alameda, California, working with AAV, she came to CHOP to manage regulatory affairs. Like the others on the team, she has a natural ease with children.

A team of regulatory project managers and reviewers helps the researchers to design the study and refine the protocol, then oversees manufacturing and quality assurance. It's critical to get the details as correct as possible, right from the start. Explains Jean Bennett, "If you want to change one thing, then you have to do the clinical trial application all over. It's expensive. Each vector prep, for example, costs five hundred thousand dollars. If you kept tweaking things to make it work better, it would never get to trial. So early on we tested a variety of different things and saw what worked well and stuck to it, knowing that the technology will improve."

After initial preclinical testing comes navigation through the various agencies, a difficult journey that became familiar to the Canavan families as three clinical trials progressed, and to Lori and Matt Sames today as they help to plan the clinical trial for gene therapy for Hannah. The path is littered with government-speak acronyms: the FDA's CBER (Center for Biologics Evaluation and Research) and NIH's RAC, then independent DSMBs (Data and Safety Monitoring Boards) and local ethics boards, biosafety committees, and additional local oversight. Finally, the IRB (Institutional Review Board) at CHOP set the doses of viral vectors for Corey's trial. "It grew so complex that Jean Bennett developed a Monopoly board to demonstrate the process. Scheduling is a nightmare," says Wellman. Pending legislation in the United States, however, will streamline clinical-trial-protocol approval. For example, as it stands now, if team members hail from five institutions, then five institutional review boards are required to give the clinical trial protocol the go-ahead. The proposed rule change, filed in July 2011, will enable one IRB to oversee a multicenter effort.

The CHOP trial required extra oversight because, unlike the other two trials—in the UK and at the University of Florida—the protocol included children. When the researchers submitted their material to the RAC—including scientific and plain English abstracts, the voluminous clinical protocol, the informed consent documents, and background on all the researchers—they knew that the inclusion of children was a "special clinical concern," because of the "not life-threatening disease and high risk gene therapy," in RAC language. The trial was chosen for presentation at the next quarterly meeting, the last of the year. So on December 13, 2005, the husband-wife team of Al Maguire and Jean Bennett, along with Fraser Wright, who'd designed the viral vector, and two sets of parents who had young children with LCA2, sat before the committee in Bethesda. They took turns addressing the concerns of the committee members, explaining that children were critical to testing the therapy because their retinas had larger islands of preserved cells that could be corrected than adults had.

Dr. Jean and Dr. Al, as Corey calls them, were passionate and convincing. The RAC recommendation was unanimously favorable, and it was on to the Institutional Review Board and Institutional Biosafety Committee, where the researchers faced even more detailed questioning about the justification for the trial, the particulars of the research design, the exact procedures, and subject selection. The matter of children was a key point. How would they explain to a child a procedure that most adults have never heard of? How would they convey the idea that the experiment might not help the child? Obtaining truly informed consent was the next hurdle. The case of Jesse Gelsinger had led to lasting changes in the process. "I think if any good is to be gained from his unbelievably sad experience, it is that it racheted up the safety and protection of clinical trial subjects in gene therapy," says Bennett.

The NIH provides extensive guidelines on how to handle communication around informed consent, for kids and adults. Their documents offer examples in seventh-grade-level language, such as "It is always okay to change your mind about being in the study," and calling the viral vector a "transportation system to deliver the gene." The NIH guide also points out what a researcher *shouldn't* say, such as "Your doctor is recommending this experimental treatment." It is this type of promising language that feeds the "therapeutic misconception" that draws the bioethicists and media to clinical trials that have tragic outcomes.

Corey, turning eight three days before the procedure, just made the age cut-off. "Seven is the age of reason, and eight is the minimum age to ask kids directly if they want to participate in an experiment. For our protocol, we needed an eight-year-old," says Kathy Marshall. Corey, Nancy, and Ethan's talking and listening to Kathy was step 1 in the clinical trial process: enrollment.

Step 2 was the informed consent. "It was a process, not a one-time thing. We discussed the whole trial with Corey, over a period of weeks. He had a document his parents read to him, and they discussed it. Questions went back and forth. He asked appropriate questions, and we knew that he understood," says Bennett.

Al Maguire was the team member who actually sat down with Corey one-on-one when it came time to sign. The surgeon seems a little gruff at first and may be a little scary initially to a small child because he is very tall, but he is screamingly funny and quickly charms children. "Dr. Al" patiently explained everything, including all the procedures and the risks for each test. An electroretinogram, for example, has a risk of corneal abrasion—but Corey was a veteran of ERGs. He also told Corey the risks of anesthesia and surgery, side effects such as high blood pressure and weight gain, and taking prednisone before and after the surgery. Corey focused on the fact that the prednisone tastes like cherries.

When Nancy and Ethan discussed the procedure with their son, they couldn't help but use more hopeful words than the NIH guidelines suggest, but they tried to contain their enthusiasm. "We told Corey that the surgery would make him see better. Dr. Al asked him what he thought, if he was scared, if he had any questions. Everybody was involved. He was scared, of course, but he's happy he did it," says Ethan, smiling.

A clinical trial participant under seventeen can provide assent, which is an affirmation that is not quite as strong as consent, which requires legal ability. "Assent is not just the absence of 'no.' We needed consent from the parents and assent from the child. It's delivered as a one-page document or verbally, and if the kid says 'no' or 'stop,' we must do so. And the informed consent can't use coercive language such as "Your mom and dad say it is okay,'" Kathy explains.

The consent form that Nancy and Ethan signed was twenty-one pages. "It had things like if infection occurs, they might have to remove the eye, then use a prosthesis, and that they're responsible for it and you get lifelong care," Ethan recalls. They didn't hesitate to sign. "I thought he might go blind if something went wrong, but that was his future anyway. Why not try it? It worked in the animals."

The Haas family arrived in Philadelphia in plenty of time for step 3 in the clinical-trial process: baseline testing. After bloodwork and scans, Corey had tests of his visual acuity, visual field, pupil responses, and dark adaptation, and answered questions about his "activities of daily living." He tried to navigate a mobility course, which went on for an excruciatingly long time as he struggled to see even the vaguest outlines of the objects strewn in his path. More ERGs confirmed the lack of electrical activity in his retinas. A video taken on September 20, two days before he turned the magical age of eight, shows Corey hesitantly walking down a hospital corridor holding Nancy's hand, tapping a cane.

Step 4 is administration of the treatment. The trial was a dose-escalation design, similar to the one Jesse Gelsinger took part in nine years earlier in the same city. Groups of three individuals would each receive progressively higher doses of the gene-laden vector, including five children, aged eight through eleven, all of whom had some vision. The dose would go up every six weeks when three new people began, once things got rolling, leaving plenty of time to see any adverse effects. Each time the dose increased, a data safety monitoring board would meet to give a thumbs-up. "The strategy was simple: low, medium, and high dose. Corey was medium. Dr. Maguire determined where to place the injection in the macula based on where each patient's retina was the most complete," explains Dr. Jean.

Late on the morning of September 25, the Haases walked the few short blocks from the hotel to the hospital, and sat for a while, waiting. Corey was neatly dressed in khaki pants and a greenish T-shirt with a gaping lion's mouth, clutching his silver-tipped cane and squinting at his handheld Nintendo, positioned right against his face. He seemed to share the bravado of the lion on his shirt until a nurse came to check his vitals, when fear started to creep into his eyes. Nancy and Ethan managed to keep him smiling, but then it was time to change into hospital attire. Corey's pajamas were covered with astronauts, planets, and aliens, repeating the celestial theme of the carpeting at CHOP, the cartoon spacemen oddly echoing what the doctors looked like in their blue scrubs and masks and white hairnets. When a nurse leaned in to gently mark a red X above Corey's left eye, his terror became palpable, and when she gave him eyedrops, they spilled over into a torrent of tears. Abandoning his bravery, Corey crawled onto Nancy's lap, and their tears mixed as Ethan leaned over to comfort his family.

Soon Corey was given a sedative, then taken to the operating room. As the general anesthesia knocked him out, nurses draped him from the neck down in a white blanket with three bold blue stripes, and he was belted to the table at his midsection. The tubes from the anesthesia snaked from his mouth. A technician taped Coreys' right eye shut, swept Betadine over the surgical field on the left, and affixed tiny metal prongs to hold the eye open. To the side, on a table, sat the equipment for the gene transfer, including two syringes full of the loaded viruses. From the ceiling, an ophthalmoscope with several eyepieces was lowered and hovered, like a strange mechanical spider, and soon a magnified image of Corey's prepped left eye appeared on a large computer monitor to one side of the table. Finally, all was ready.

Dr. Al, ophthalmoscope in place, leaned in and cut open a tiny flap of tissue in Corey's left eye. Delicately, he removed the clear, jellylike vitreous humor that fills out much of the eyeball and replaced it with air. The jelly would reform on its own. He then introduced about 48 billion adeno-associated viruses, each harboring a healthy copy of the *RPE65* gene, near the retinal pigment epithelium, the cell row unable to make the RPE65 protein that enables the eye to harness vitamin A. All went well, except for a tiny rip in the retina that Dr. Al repaired with a laser.

Corey woke up fourteen hours after he went under to find his worried parents on either side of him. He felt okay. "I hated having to lie on my back and hold still," he recalls. He had a runny nose from all the eyedrops, and the eagerly anticipated cherry-flavored prednisone upset his stomach. He had a swollen head for three days, and the white of his treated eye was red for a day—but that was about it for aftereffects. He was able to visit the Liberty Bell just a few days later, still using his cane.

Drs. Jean and Al had been fairly confident that the gene therapy would work, having taken home the various canines who'd had their vision restored. But they could hardly believe what happened at the zoo, how fast it was, just four days after the gene therapy, when the sunshine was painfully bright to Corey, the boy who'd spent his infancy staring into lit bulbs. His response was right in line with the young adults who'd had the gene therapy in the various clinical trials. Looking back, Dr. Jean relates the first signs of vision in the various patients. "Corey can walk to his friend's house at dusk and find things he drops. Other kids have said, 'I can see stars!' or 'Now I can play soccer and don't need someone next to me to help me find the ball.' A forty-four-year-old who was nearly completely blind can now tell when her children need haircuts. It's all very emotional," says Dr. Jean.

A whole new world opened up for Corey, and it took some adjustment. "Corey's reading skills dipped a little at first, because his eyes wandered at all the new words showing up in his expanded visual field," recalls Ethan. Corey saw his friends' faces clearly for the first time. "He asked me when two girls had dyed their hair blond," recalls Nancy. "They were only seven years old. But their hair had appeared dark to him before the surgery. Before, he recognized people by their voices." Now he could see them.

Corey Haas became the boy who saved gene therapy in a figurative sense, and also in a real sense. Gene therapy was a field that desperately needed a happy story, and the boy with the failing vision fit the bill. Corey didn't have an inexorable brain disease like Lorenzo or Oliver or Max. He didn't have failing immunity like Ashi and the leukemia boys, nor did he face a race against time to stall or reverse paralysis, like Hannah. He simply couldn't see very well, and wouldn't have been able to see at all once he passed adolescence. And the isolated nature of his problem, in the eye, beyond the surveillance of the immune system, made it unlikely that he could suffer the fate of Jesse Gelsinger.

Corey wasn't the first to have the successful gene therapy for LCA2, but, for a time, he was by far the youngest, and gene therapy is an intervention that will one day routinely be done *before* symptoms appear. The LCA2 gene therapy story is also one of cooperating and complementing scientists, pioneering female researchers, and animal research in which the doggies get to go home with the children that they help.

Follow-up to gene therapy is as important as the procedure itself. Step 5 is to repeat the baseline tests to track what's improved, and step 6 is reporting the results. Finally, step 7 is to evaluate the results over time. (This is what eventually fell by the wayside in the case of Ashi DeSilva, who received the very first gene therapy in 1990. She's alive and well, married and happy, and that would seem to be all the evidence needed to consider her gene therapy a success. But French Anderson, who led the study, is still trying to locate and publish the long-term follow-up data.)

The clinical trial in which Corey was treated ended in the summer of 2009. By late fall, the research team had already submitted a phase 3 protocol to the FDA, RAC, and DSMB. In addition to expanding the number of patients treated for LCA2, this trial would include three-year-olds. Given the startling success at the zoo, and in the other patients, the phase 3 protocol easily passed the RAC by year's end, while plans were in the works for treating the second eye of some of the patients.

A trip to the zoo in bright sunshine four days after gene therapy was enough to convince Corey and his parents that the gene therapy had worked, but the researchers had to delay publishing a paper, to see if the effect persisted. That paper appeared in the prestigious UK journal *The Lancet* on November 7, 2009, concluding, "an 8-year-old child had nearly the same level of light sensitivity as that in age-matched normal-sighted individuals." Gene therapy had transformed an almost blind child to almost normal.

The PR machine at Penn and CHOP trumpeted the news. *The Wall Street Journal, The New York Times*, the *LA Times*, Reuters, ABC, CBS, CNN, and NBC all covered it, as of course did *The Philadelphia Inquirer*, the newspaper that a decade earlier had vilified gene therapy in the wake of Jesse Gelsinger's death.

The informed consent documents that Ethan and Nancy signed hadn't prepared them for fame, but that's what the family encountered when they were whisked down to New York City in the wake of the *Lancet* paper, an environment as different from the foothills of the Adirondacks as is possible. Corey recalls, with a grimace, having to go shopping for shirts, ties, and dress pants.

"Hey, aren't you the kid who was just on TV?" a man asked Corey at the Empire State Building. The family had indeed just been on *Good Morning America*. The interviewer, like others, assumed that Corey had been completely blind and had overstated the gene therapy success, but Corey was savvy enough not to contradict her. The family appeared on ABC and CBS (although thanks to a Jets game, they never saw themselves on the latter).

On the train ride back upstate, Ethan fielded calls from reporters. "I talked to the *Today* show, Bloomberg News, *Marie Claire* magazine. We heard from Japan and Russia, and in China we were on the front page." Oddly, the Haases' hometown newspaper, the *Post Star*, interviewed the family but held the story until the end of November, when it ran on the front page.

Corey took his instant celebrity in stride. He returned to school, and the family began to look forward to getting his second eye fixed. But first the researchers had to monitor Corey's progress and assess it at the two-year mark: September 2010.

19

BACK TO CHOP

ON A SUMMERY WEDNESDAY AFTERNOON IN MID-September 2010, I entered the atrium of Children's Hospital of Philadelphia, one of several behemoth medical buildings on the southern edge of the ivy-covered campus of the University of Pennsylvania. The clangs of construction mingled with the horns of cars stuck in the ever-present traffic as a helicopter whirred above, spiraling in to land on the roof. Founded in 1855, CHOP was modeled after Great Ormond Street Hospital for Sick Children in London, and currently serves more than a million children a year. "Hope Lives Here" is its motto.

The first-floor lobby is a bright, open space, with giant-sized gizmos and gadgets for kids to look at and operate. It seemed to me more like a science center than a hospital. A wide, open staircase leading up passes vibrant murals on the walls. Where the clear plastic steps end on the second floor, a celestially marked carpet begins, and children can follow the colorful stars and spaceships in several directions. I followed signs to the Wood Building, where Corey was having tests to evaluate his progress two years after his gene therapy. Along the way, posters on the walls told the stories of children.

"Lia's mother gave life to her. Twice," described a partial liver transplant.

"Maddy not only beat bone cancer, she walked away from it," was accompanied by a photo of a girl with short brown hair, hugging a dog.

"Something Hunter never thought he'd see in his lifetime: a lifetime," referred to a stem cell transplant.

As you turn into a walkway approaching the Wood Building, the photos were suddenly hugely blown up: the face of a child with Down syndrome, the tiny hand of a preemie, the distinctive features of a child with Cornelia de Lange syndrome. Kids hurried along with their families. One child had an eyepatch, another sported goo on his head from electrodes used to measure his brainwaves. A few children zipped by in wheelchairs, their parents straggling behind. New parents schlepped bags stuffed with typical baby and medical gear.

I tried to guess at the diagnoses. Did the newborn with both arms in a sling have osteogenesis imperfecta, the brittle bone disease that sometimes gets parents wrongly accused of abuse? Did the seven-foot-tall young woman with long limbs and giant hands have Marfan syndrome, the condition Abe Lincoln is thought to have had? As I found my way to the ophthalmology department, I saw an adorable little boy with a big head in the waiting room—hydrocephaly? The waiting room was packed with young patients and their parents. Some adults sat quietly, wiping away tears, while others squeezed into the tiny chairs to help their children work puzzles. A huge man cradled a tiny baby who was wearing an eyepatch. Most of the kids were clad in shorts and tank tops that suited the hot and humid Indian summer day.

Mid-September was a strange time to visit CHOP, for that was when, eleven years ago, Jesse Gelsinger died, four days after his gene therapy for OTC deficiency. Two years ago, Corey had his gene therapy here, and four days later, he had responded to light for the first time. The two-year tests that he was here to take, mandated by the clinical trial protocol, would measure his vision at several levels, from navigating a mobility course down to the molecules and cells that provide sight. But Corey and his parents, Nancy and Ethan, hardly needed tests and scans to tell them that the gene therapy worked. They witnessed it every day.

The PR person ushered me through a door that opened onto a long, narrow corridor to exam room 1073, home base for the next three days. Here Kathy Marshall, clinical research coordinator for the gene therapy trial, was taking Corey's vital signs. She's slim and slight with short dark hair, a grandmother who doesn't look the part. Corey, comfortably clad in jeans and a purple-and-green T-shirt that matched his sneakers, occupied a big exam chair at the center of the small room, while Nancy and Ethan sat off to the side. It was Corey's tenth birthday, and Kathy had brought homemade chocolate cupcakes that looked like they were about 90 percent frosting.

Kathy, an ophthalmologic technician, ran the show. The first test was a favorite—visual acuity. Corey faced a large white board that had rows of circles that opened like windows, revealing a letter behind each. He wore a blue patch with stars and spaceships on first one eye, then the other, and read off the letters as Kathy opened the little windows. The advantage of this variation of the standard eye chart is its flexibility—Kathy easily slipped large sheets of letters behind the board. "We have to switch the charts because most kids memorize where the letters are. We don't do it straight across for that reason, too, we mix it up," she explained to me. The letters shrunk as Corey read lower on the board, and by the bottom row, he couldn't see them.

Right behind the big eye chart a shelf housed a stuffed elephant, a dog, and a duck, each perched in a cubby with a little red light beside it. "Those are to test visual acuity in children too young to read the letters on an eye chart," Kathy said as I reached for one. She turned back to Corey. "Time for the Farnsworth test. No dilation today."

"Yay!" Corey jumped off his throne and took a seat in front of what looked like a large box of eye-shadow containers. This test *did* look like fun. He used a magnetic stick to move the circle of color toward a test colored disc in the lower right, ordering the colors from most to least like the test circle. A few years ago, he couldn't even see the discs. Next he took a classic test for color blindness, searching for embedded patterns in a landscape of Xs and Os.

Next came the prep for pupillometry. This test would show how far Corey had come from his "owl eyes" days, when his pupils wouldn't contract even when he stared straight at lit bulbs. Because Kathy hadn't given him dilating drops, he had to sit in complete darkness for a half hour, wearing two layers of eye patches, to coax the muscles around his pupils to contract as much as possible, opening the aperture. Corey dislikes the "dark adapt" period, but not as much as he hates eyedrops. Both interventions have always been part of his life, so the various doctors could see what was going on, or not going on, in his eyes.

Outside room 1073, Jean Bennett was running back and forth, setting up the test in the room across the corridor. She wore her trademark floral jumper with black tights and top. When all was ready, Kathy led Corey into the darkened room and sat him before a device that resembled a microscope. The patches came off, and he looked into the eyepiece as Dr. Jean watched a computer screen with a red filter over it. I squeezed into the room with Nancy and Ethan, and Kathy shut the door. A light flashed in front of Corey.

"Perfect!" said Dr. Jean, staring intently at the screen. "Good good good good GOOD!"

We all leaned forward to see the giant image of Corey's pupil suddenly constricting, like a guppy's mouth snapping shut. We watched as lights of different frequencies flashed before Corey's eyes and his pupils instantly responded. "You're in the home stretch!" yelled Dr. Jean, but then her voice lowered. "When he took the same test two years ago, before the gene therapy, he had no response at all."

Pupillometry done, Dr. Jean updated Corey on Mercury, the Briard mix he'd met in July, currently residing with her and Dr. Al Maguire. Mercury and two other dogs were awaiting permanent homes with LCA2 families. Mercury's mother, Venus, had just that morning peed in the kitchen. As we all chatted about our pets, Dr. Jean wheeled Corey back to exam room 1073, now adorned with colorful balloons. Corey demolished a cupcake and wiped frosting from his face as Kathy began to prep him for the contrast tests.

These are fast. One eye at a time, Corey read shaded letters on a chart, the darkest in the upper left, blank ones at the lower right. Kathy noted the points at which he could no longer see the letters.

Soon we were all trooping down the hall, single file. The doors to some of the exam rooms were open, their patched or bespectacled occupants waiting, some oblivious and exploring, others trying to escape or wailing. A little girl with red ringlets toddled into the hallway as her parents tried to soothe her shrieking older brother.

Kathy guided us into the last room on the right, shut the door and lights, and sat Corey in front of a half globe that looked like a mini-planetarium. Corey peered into an aperture in the inside face of the globe, while Kathy settled herself in a seat behind the hole, watching Corey's eyes through a telescope. Nancy leaned in to help Corey patch one eye, and then a mechanical arm with a light at its tip moved from behind his head, first around from the right, then from the left. Whenever the light came into his field of vision, Corey pressed a button. Then Nancy switched the patch to Corey's other eye, and the test continued. Corey was having fun. For we observers, it was about as exciting as watching paint dry, but this was a critical test that revealed what Corey's brain did with the new visual input. "After the injection you can see a very large increase in the visual field. In contrast, in the uninjected eye over time, we see the visual field deteriorating," Dr. Bennett explained when she joined us as Corey finished the test.

Corey seemed unusually at ease with the medical staff. As I watched their interactions, it was easy to see why. These are very special people. At the core of the team are Drs. Jean and Al.

Jean Bennett and Al Maguire fell in love over a dissected head. They were in the neuroanatomy lab during their first year at Harvard Medical School, and Jean, like Corey would years later, became captivated by Al's droll sense of humor. He spoke in the same deeply resonating voice when naming the regions of the exposed brain between them or when cracking a joke. "We bonded over the hypothalamus," she recalls with a smile.

They married, and three years later, in 1986, when they received their medical degrees, they also had the first of their three children. Their career paths converged when they treated Lancelot and the other dogs for LCA2. Jean, who had a PhD from the University of California, Berkeley, in cell and developmental biology, had spent a year at the National Institutes of Health, where she met some of the founders of gene therapy. It was French Anderson, who would go on to do the very first gene transfer experiment on four-year-old Ashi DeSilva in 1990, who noticed the gifted and enthusiastic young woman and encouraged her to go to medical school. The idea appealed to her. "I knew that I loved research, but I wanted to gear it toward diseases where there was a need for developing better treatment."

Bennett became expert at engineering mice to carry human disease genes, which were in high demand. She helped a researcher who was working with a mutation that affects the eyes, and became intensely interested in the visual system. "I applied for a career development award from the Retinitis Pigmentosa Foundation, which is now called the Foundation Fighting Blindness, and I got it." She continued her genetic research in a small lab she set up in the Michigan hospital where Al completed a fellowship in 1991, focusing on retinal degeneration. Then in 1992 they both went to Penn, thinking about one day doing gene therapy on the eye. "Even though it was still fantasy, it was becoming more and more realistic," Bennett says.

Of all the people that Corey, Ethan, and Nancy interact with, they seem closest to Dr. Jean. The sprightly physician-scientist grew up near Yale University, where her father, William, was a professor of physics, and quite a phenomenon. He discovered and developed several types of medical lasers. Father and daughter even co-invented a laser device to detect and describe heart murmurs.

William Bennett followed the LCA2 gene therapy progress

with great excitement, but he died just before Corey learned that he'd become patient number 9. "He was so fabulous," Dr. Jean says, her eyes shining. "At Yale he was the master of a residential college, so he lived with students. He had two who were blind, who used canes and guide dogs. When we started the trial, he wondered if one of them, Lauren, had the disease. Two years later, Lauren called and told us that she had the *RPE65* mutation." Since the researchers have stockpiled enough healing virus to treat everyone with Corey and Lauren's disease, the young woman will likely be able to regain her eyesight.

Corey was in remarkably good spirits that first afternoon, and the next morning he was fine, too, for his 8:00 a.m. appointment with an fMRI machine. This test would reveal the parts of the eye that are most sensitive to light. It is particularly important for following up the older people who had the gene therapy, who had fewer photoreceptors to salvage than did Corey. "fMRI demonstrates unequivocally that an optic pathway is intact and reactivated, even after thirty-five years of degeneration," says Bennett.

The next test—the mobility course—is Corey's favorite, the one replayed on YouTube and shown to Congress. He looked on as Dr. Jean darted about, placing squares of carpet on the floor of the room across the hall. Then she laid down black felt arrows on some of the squares, creating a pathway. A few squares rose several inches, to test Corey's depth perception. Once the squares were set, Dr. Jean hauled in a lamp on a tripod to serve as an obstacle, as Kathy placed a few foam cups along the path and then suspended a piece of foam pool noodle horizontally. It looked like they were setting up for a stage production.

"That test is fun! I just follow the arrows," Corey told me, grinning, as he twirled on a stool.

"Quit fooling around!" said Ethan. He got up and led Corey to the other room, where Kathy approached with a patch for his right eye, the untreated one. Dr. Jean gave the instructions: "Just follow the arrows. We'll score the number of objects you bump into. Remember to look up and down, too. Some things you have to step over. The light will be dim. When you're done, just open the little door, under the hanging stop sign." Corey fidgeted; he knew the drill. At the start signal, he jumped up and navigated the pathway, perfectly, in about fifteen seconds. As he expected, everyone clapped.

Corey returned to exam room 1073, while the adults closed the door and quickly switched around the pathway, because kids memorize that, too. Kathy came out and stopped Corey's spinning on the stool, then switched the patch to his treated eye. Corey's demeanor changed immediately.

"I can't see you, Mom," he said to Nancy, who sat about three feet away. He could just barely make out the printed flowers on her shirt, and reached out to touch them.

Kathy guided Corey across the hallway. Once he stood in front of the mobility course, the difference between his two eyes was instantly obvious, just from his body language.

Dr. Jean led him to the first square. He extended his left foot, tentatively, then pulled back. "I can't see anything," he said in a quivering voice. His shoulders slumped. But then he pulled himself together and tried, slowly, to walk. When he was about to crash into the lamp, Dr. Jean intervened. It was already long past the time it took to complete the course using his good eye. He turned, took a step up, then down, and landed at a black square, one without an arrow. He was off course.

"I have no idea where I'm going," Corey said as Dr. Jean reached out to keep him from walking straight into the wall. Then he hit a foam cup and crashed into the suspended pool noodle. He stepped off the path. "I still can't see anything." He was facing 90 degrees from the three-foot-tall door that marked the end of the course, so Dr. Jean gently turned him. But rather than attempt to walk, he instinctively reached out and felt for the door.

Corey recovered his confidence quickly when Dr. Jean asked him to walk the course with both eyes open. Showing off, he did it backwards. But then she gently persuaded him to do it with his bad eye again—to get more data—and he fared as poorly as he did before. This was what Corey's vision would now be if not for the gene therapy.

The mobility course is an example of how scientists must sometimes invent creative ways to assess experimental outcomes. For example, since chickens won't take eye tests, Susan Semple-Rowland entices her subjects, cured of LCA1 with gene therapy, to peck at Skittles candies to demonstrate their visual abilities. The spinning dogs taught their masters how to measure success. The makeshift mobility course is similarly low-tech—and that led to objections when the researchers were trying to push through the gene therapy protocol with the FDA. "Some people said, 'Get rid of the mobility test! It's unscientific! It's psychological, about learning and coordination!' not like a test that measures the precise number of photons hitting a surface in a given period of time," gushed Al Maguire. Then he paused and shook his head. "How can mobility not be meaningful in the life of the patient?" Ethan put it more succinctly. "He didn't move on the mobility course before, and now he does."

Corey's testing continued. Nystagmus was next, the oscillating eyeballs that were so unsettling to Nancy when she'd stare into her newborn's eyes. Back in room 1073, Dan Chung, a pediatric ophthalmologist, captured the dampened jiggly movements with a camera, then switched seats with Corey to let him do the examining. It was hard to get Corey to stop playing doctor, but Kathy was ready with a long questionnaire, part of the clinical trial protocol.

"OK, let's start. Corey, do you spend much time worrying about your eyesight?" she read from the script.

"Some."

"Do you have difficulty reading ordinary print? Like in newspapers and magazines?" Corey was momentarily stumped. "I don't read newspapers and magazines." Kathy looked up and smiled, then returned to her list.

"Do you have difficulty seeing up close, like in cooking? Or doing a model airplane?"

Corey looked over at Ethan, puzzled. "A what? I guess so."

The next series of questions were multiple choice, including the answer "moderate." Corey latched on to this new word and used it indiscriminately to answer more questions. Then Kathy, trying to keep a straight face, asked, "Can you pick out and match your clothing?"

Corey gave a classic eye roll. "I don't do that! I don't care if they match. I pick them out at random."

Kathy needed an answer, so she attempted to rephrase the question. "Could you pick out clothes that match?"

"But why would anyone care about that?" A long pause. "Okay, no."

Kathy then reasked it all, with the questions framed anew as fill-in-the-blanks. Thus is the nature of rating scales. They are stultifyingly boring. Ethan, Nancy, and I were nodding off, while Corey became increasingly frustrated.

Kathy continued. "Corey, on a scale of zero to ten where zero is death and ten is feeling good, how would you fill in the following sentence: 'I feel _____ most of the time because of my eyesight.' "

"If I said zero, I wouldn't be here. I'd be dead," he pointed out helpfully.

Next, Corey had to assign a numerical value to a long list of activities, which included reading fine print on a medicine bottle, ensuring that bills are accurate, and putting on makeup.

Finally, Kathy asked, "Are you irritable because of your eyesight?"

Corey looked confused again. "I don't know what that means."

"Grumpy."

"No. I'm opposite that." A big grin.

Four hours, four tests, and it was time for lunch.

Returning to the testing area in the ophthalmology department, I walked past the very tall woman who might have had Marfan syndrome, a teen with only two limbs, and a pair of Asian toddler twins. A few families had joined the waiting-room population, which ebbed and flowed like a tide as the children were taken back to the dreaded warren of exam rooms and new arrivals took their places.

Corey was sitting in his chair at the center of exam room 1073, raring to go. It was time for another visual acuity test. "I like to do it two or three times a visit. Sometimes people see better at a certain time of day," Kathy explained. "And we're getting three times the data!" chimed in Corey as Kathy slid the big plastic sheets of letters behind the holes.

Soon it was on to the prep for the OCT, optical coherence tomography. This was the imaging technology used on LCA2 patients in 2005 that revealed that they had residual rods and cones in good working order, making the gene therapy possible. "Do I have to dark adapt?" Corey asked Kathy.

"Nope."

"Good!" After a few dark adapts, he didn't mind using dilating drops instead. Kathy gently tipped his head back and quickly dripped in numbing drops and then dilating drops. "It tickles!" he giggled, and then his grin turned downward, his feet flipped back and forth, and he emitted a long "ow" as the drops' delayed sting set in.

Kathy escorted him to a room at the end of the corridor opposite where he took the visual field test and sat him before another microscope-like contraption. She turned the lights out as Nancy helped Corey get his head into position. The hardest part was getting him to stop talking. The test began. Corey saw a flash of red light, and then columns of color, like multihued ephemeral stalagmites, flickered onto the monitor.

"Look up. Now down. Now to either side," Kathy directed as the software translated the stalagmites into six dazzling, multilayered cross sections of Corey's retina. On the screen, about two-thirds of the way down the retinal sandwich was the all-important RPE, the cell layer that lacks the protein that would enable Corey's eyes to use vitamin A. Fittingly, the software assigns the RPE the color of a carrot—orange.

Kathy scrutinized the colored layers on the screen, searching for the precise spot where Al Maguire injected healthy *RPE65* genes two years ago. It could be the area with a lot of orange stuff. She carefully cropped the image to capture the purported injection site as Corey came around to look. "Is it time for fundus photos?" he asked excitedly. "Am I gonna see Sonia?" Kathy nodded a yes.

He swung himself one seat over, behind the Soma TRC 50EX Topcon retinal camera, and knew exactly where to put his eyes and chin. A cheerful, willowy Asian woman glided in and sat opposite Corey and adjusted the device. Corey's fundus, the back of the eye, suddenly filled the screen in front of Sonia.

The image looked like Mars. Corey's right fundus appeared as a reddish orb with canal-like blood vessels snaking out from a bright white spot, the optic disc. His left eye had a giant, ragged white starburst near the white spot, marking the site where, two years ago, Al Maguire delicately placed the virus-clad genes.

Fundus photos done, we headed back to the exam room and awaited Dr. Al, who was rushing over from a crowded clinic. He's tall and commanding, with long gray hair and a very deep voice. He sat on the stool, slipped an ophthalmoscope over his face, and wheeled over to his famous young patient. While he aimed the lit optical device, which looked like a Darth Vader mask, at Corey's retina, Dr. Al reported on the antics of the pets in the Maguire-Bennett household. "And I had to keep Jean from playing with your birthday present," he joked as he glanced at the huge wrapped box in the corner, which contained a radio-controlled Rock Crawler jeep. The doctor looked deeply into Corey's left eye and asked him to look in all directions. "Acuities are symmetric today. Very interesting," he said to Kathy, then sat back.

"Corey, did Dr. Jean talk to you about eye number two?" asked the doctor.

Kathy straightened, looking uncomfortable. "What?" asked a puzzled Dr. Al. "Am I not supposed to ask that?"

Kathy glanced over at Ethan and Nancy, then said with an uneasy half-smile, "No." The last documents to the FDA were being submitted that very day, and the team wouldn't get final word on the next clinical trial for several more weeks. It was too early to mention, but Dr. Al wasn't picking up on Kathy's body language. The parents did. They leaned forward, because this is what they had come to hear—if and when Corey's second eye would be treated.

The doctor turned back to the patient, slightly flustered, and examined each eye closely. "There may be, Kathy, will be . . . a day . . . can't say when. FDA once we get through the IRB we can go forward and we talked . . ." He wasn't making sense. Kathy gave him another warning look and he stopped talking about the second eye.

"Crystal clear times two," he concluded, wheeling backwards and leaning down to store the instrument in a cabinet.

"Wow, that was fast!" said Corey.

"After a hundred billion times doing this, you get pretty quick."

Next Dr. Al showed Corey a device to measure the pressure in his eyes. His voice lowered into a robotic sound. "Do you know what this is? It's Hal. And my name is Dave."

Corey didn't recognize HAL 9000, the computer, or Dave Bowman, the lost astronaut, from the film *2001: A Space Odyssey*, but Kathy and I smiled as the doctor expounded upon the intellectual value of science fiction. Then he abruptly turned to Kathy and extended an open palm. "Hand me the knife!" he boomed.

Instead of wielding a weapon, Dr. Al applied his "closed eye technique" to drip drops into Corey's eyes so he could measure the pressure. He turned to the rest of us. "My greatest contribution to medicine is proper administration technique." Corey squirmed a little.

"Settle down or we'll make you have another MRI," Dr. Al bellowed, sending Corey into a fit of laughter. To calm the boy down, the doctor asked how school was going. When Corey didn't seem too thrilled, Dr. Al said, "Should I write you a note that excuses you for the next year?"

Corey broke into more giggles, but stopped soon because he knew it was time to sit still so Dr. Al could give him more numbing drops for the coming ERG test.

"Why do I need numbing drops to prevent the sting of the dilating drops, when the numbing drops sting too?" Corey asked.

"There's no negotiating with terrorists," said Dr. Al as he stuck patches decorated with yellow trains onto the boy's eyes. He flipped off the lights, exited, and left Corey, Nancy, and Ethan for another dark adapt period.

"Corey hates this," warned Ethan forty minutes later as we piled into the room at the far end of the corridor, where the visual fields were done yesterday. This was the electroretinogram, the dreaded, invasive ERG. "Years ago, in Boston, the ERG took a half hour to dark adapt, then two hours of crying and screaming," he added. Ethan pointed to what looked like one of the curtained photo booths that used to be in shopping malls before everyone had camera phones. "That's for the younger kids. Corey used to go in there."

But Corey now used the grown-up device. There wasn't a peep out of him as Sonia administered lubricating drops and then placed tiny cups that resembled contact lenses directly onto his numbed eyeballs. Lanyard-like cables snaked out of the eye cups and an electrode dotted his forehead. If Dr. Al was Hal 9000, then Corey was surely a Borg, from *Star Trek: Next Generation*.

"Wow, that was fast!" Corey didn't seem afraid at all.

"You really grew up!" said Sonia, who last tested him six months ago.

"Yes, now I am mature," Corey said, smiling up at her.

Kathy turned the lights completely off, and the only illumination was from a computer monitor, three feet or so from Corey's left side, draped with a red gel sheet.

"Are you ready?" asked Kathy, standing at the monitor.

"Yup!" Corey looked into yet another set of eyepieces.

Flashes and flutters of light appeared, and Corey pushed a button when he saw something. Over on the computer screen, red and green lines flickered and pulsated into life, as the software compensated for Corey's jiggly eyeballs. The patterns of the ERG were confusing, because the left eye showed up on the right side of the screen and vice versa. But even I could tell, from having seen

Corey's flat-as-an-Iowa-cornfield ERG that nailed his diagnosis years ago in Boston, that he was seeing.

When the flashes and flutters finished, Corey backed away, Sonia removed the eye cups, and he went to stand next to Kathy at a second computer, where the Mars fundus was displayed to the right of the ERG squiggles. "This is fun!" Corey chimed, a little too cheerily. "Whatja get?" Kathy explained the colors on the screen.

ERG done, Thursday's testing ended. As we walked down the hallway, through the now empty waiting room and toward the reception desk, a woman stared at Corey, then rushed over to introduce herself. "I know you from TV!" she purred. Corey's a veteran of the local news.

Descending the stairs to the atrium, he looked around at the beautiful murals, displays, and gizmos, and grinned. "I really like it here."

"Not every kid here would say that," said Ethan, but Corey didn't hear him.

When my youngest daughter, Carly, was a toddler and about to throw a tantrum, we'd call her a thundercloud. Friday morning, that was Corey.

I arrived at exam room 1073 late and flustered, afraid I'd missed something. Nancy and Ethan were there, more subdued than during the past two days. Corey was in the bathroom. A few minutes later, he came in, head down, shoulders hunched, silent. Something was wrong.

"He has an upset stomach," said Nancy, pulling him to her.

"He's worked up over the next test," Ethan whispered.

Kathy, a little uneasy, joked about the original consent form for the clinical trial. "We had to ask Corey not to have kids for the next year."

"Ow, my stomach hurts," Corey cried, grabbing his middle, starting to tear up. Nancy ducked out in search of ginger ale.

Yesterday, Corey hadn't cried over the stinging eyedrops, the

ERG eye cups, or the monotonous dark adapts. But it can be hard to know what will disturb a child. "The last time, six months ago, he got upset at this test when he couldn't see the lights, because he didn't realize that that is part of the test. Sometimes the lights aren't there. He felt like he was failing a test, like in school," Kathy said in a low voice, then turned to Corey. "It's okay to say the light isn't there," she reassured him. But still, he seemed agitated. Part of the problem was physical. Because of Corey's extreme nearsightedness, he occasionally sees flashes of light. He was worried that he wouldn't be able to tell the test flashes from the unprovoked ones.

The procedure Kathy thought he was worried about, the full-field sensitivity threshold test, was next. Corey would hear a series of beeps, some of them followed by flashing lights of different intensities, and he would push a button when he saw something. The test was tedious, in six parts—four in white light, one in red, and one in blue.

Nancy returned with the ginger ale. Corey took a few sips, then grabbed his middle, and the tears flowed as he clung to Nancy. It took a lot of coaxing, and time, to get him to walk down the hallway for the test. But eventually he went, head down, and settled into the seat in front of yet another device. "These are just guesses," he muttered, halfheartedly pressing the button.

Kathy turned around and hunched over her cell phone to call Dr. Jean, who was waiting at her lab to perform a new type of scan. Kathy listened a few minutes, put her phone away, then said, "Okay, Corey, Dr. Jean says we can skip the blue part."

Corey perked up instantly and seemed his old chatterbox self as we returned to the exam room, where he settled into Nancy's lap. Unfortunately, Kathy needed to put dilating drops in again, because the flood of tears had washed the earlier ones away. As Corey scrunched his eyes closed, Kathy quietly picked up her cell phone and hit a key. "We're an hour behind," she whispered.

Next we left the department and headed up to the top floor, then navigated twists and turns until we reached Kathy's office, for confocal laser ophthalmoscopy. Corey calls it "the Heidelberg test" for the brand name on the instrument. This scan combines laser imaging and 3D analysis in real time, revealing the retina in cross section. It's essentially a fancy version of the OCT test from yesterday. Corey hates it.

"This is the test I didn't want to do!" he whimpered, clinging to Nancy, who seated him behind what seemed like the zillionth device and tried to get him to sit still. Even though most of these tests don't hurt, the constant probing and evaluating were beginning to take a toll.

"Keep it up! I'm so proud of you!" said Kathy; then she added in a low voice, "Focusing is the hardest part. The nystagmus makes him wiggle, and the machine can't totally compensate."

"My stomach hurts!" Corey now sobbed. "I'm worried about the next part!" Then he tipped backwards, nearly toppling.

Office workers nearby could hear Corey, and a few of them left. They weren't used to hearing children up here, far from the treatment areas of the hospital. I stood just outside the room, immersing myself in the arresting sketches along the walls that depict splayed eyeballs as landscapes. Corey's objections reached a screeching crescendo as Kathy reached a breaking point.

"Okay, forget about it, I'll do another part," Kathy said, scribbling some notes down. "But Dr. Dan needs a photo for an article," she added. I flashed back to a dentist who once shoved a plate full of goop into my daughter's mouth to take an impression for a periodontal class assignment, causing her misery for what to me was no good reason. But this wasn't homework. It was a clinical trial, and Corey had signed on for years of follow-up that would see him into adulthood.

Kathy tried another approach as the crying continued. "The right eye is important because it is the next for surgery—if you want to have it done. Dr. Jean is waiting."

Even Nancy was getting rattled. *"Corey Douglas. You have to do it!"*

"The FDA is waiting. We are under pressure," added Kathy.

Then the ultimate insult flew from the outraged ten-year-old, at a decibel level that reached the office ladies huddled outside. *"I'd rather be in school! NO! NO! NO!"*

Ethan leaned out of Kathy's office and explained what, exactly, the test would do to his son, who sounded as if he were being tortured. "The light will be very bright and the drops will hurt and afterward, his visual field will appear pink."

In the office, Kathy stopped to collect herself. She started over, in a calm, gentle voice.

"What don't you like about the test, Corey?"

"The bright light."

"Please put your chin up here on the rest," Kathy said quietly.

"NO!" he shouted.

But then he suddenly deflated, giving in, and a three-dimensional section of his retina materialized on Kathy's screen.

After lunch we assembled for the last time in exam room 1073. Corey had just one more test to go, which would take about two hours and had been postponed from the morning. It was Friday and Dr. Jean was supposed to have been en route to Massachusetts by now to attend a wedding, but as soon as Kathy started reporting delays, she changed her plans.

Jean Bennett's office is in a much older building about two blocks away that was once part of the Pennsylvania Hospital, the nation's first. We walked over in the intense sunshine of the 90-degree late-September afternoon and headed up to her office and lab on the third floor of the F. M. Kirby Center for Molecular Ophthalmology.

Clad in a jean miniskirt and a large metal belt, Dr. Jean ushered us into her small, crowded office. She settled in behind her desk, against a backdrop of three shelves stuffed with bulging loose-leaf binders, testament to the many years it has taken to get this far in treating LCA2. The walls were covered with diplomas and photos of dogs, including a silver-framed portrait of the beloved Lancelot, and the Bennett-Maguire children. She motioned for Corey to sit beside her, then took out papers. These were the informed consent forms that Corey would need to sign if Dr. Jean was going to administer a new type of test to measure the level of rhodopsin, the visual pigment in his eyes. More rhodopsin in Corey's treated left eye than in his untreated right eye would be a sure sign that the gene therapy had worked, and so his participation was crucial. But since this particular test wasn't in the original clinical trial protocol, Corey had to assent.

Dr. Jean talked directly to the boy, slowly and clearly.

"We are going to shine a bright light on one eye, and rush to a machine and take images at designated time points. We're going to use this test to tell people how much rhodopsin you have now, but didn't have before the gene therapy. We've already tested it in fifty people with normal retinas, including me. You, Corey, will be the first child tested. You don't have to participate, but if you do, you will be paid twenty dollars. You are going to see one bright flash and others not as bright. The flash will last four seconds. You can have pictures of your eyes after. We'll need a half hour to patch before."

Corey listened, showing no reaction, so Dr. Jean continued. "If you sign the assent, we'll take a few pictures to align your head. Then you can wear a pirate patch, and wait a half hour. We'll patch the second eye while we do the first, so you won't have two waits." She knew how much he hated dark adapting and sensed his reticence, so she leaned forward and took Corey's hand. "Would you like some time to talk to your parents alone about it?" He nodded.

Nancy and Ethan stayed in the office and closed the door as Dr. Jean, Kathy, and I retreated to the hallway, where we traded stories about how our grown kids were faring in the current economic climate. I don't think any of us expected what happened when Ethan emerged about ten minutes later.

"Corey's not going to do it."

Kathy was quiet, her face wiped of all expression. My mouth dropped open and then shut as I forced myself to stay out of it. This was his right. It is spelled out in the Nuremberg code, the guidelines developed in 1947 in response to the Nazi atrocities, to protect human subjects in experiments: "During the course of the experiment the human subject should be at liberty to bring the experiment to an end if he has reached the physical or mental state where continuation of the experiment seems to him to be impossible."

Dr. Jean didn't show even the tiniest sign of distress. She just kept smiling, put an arm around Corey, and offered to show him the equipment for the new rhodopsin test. Maybe she was hoping he'd change his mind or she could at least reawaken his curiosity, but there wasn't a hint of disappointment or coercion in her tone. Perhaps Corey did feel a little bad about refusing this kind and brilliant woman who is largely responsible for saving his eyesight.

"I'll do it when I'm eleven," he said in a hushed voice, head down. (He did.)

Dr. Jean showed him the new instrument, took us on a tour of her lab, and the three-day visit was over. Just like that. Everyone hugged.

Outside, Corey was a thundercloud again, the way the day began. Nancy was quiet, her arm around her son. Kathy was flustered. We crossed the street and walked over to CHOP. Kathy bent down and tried to make eye contact with Corey, who was examining his sneakers.

"Corey, why did you say no?"

"I don't know," he said quietly, on the verge of tears.

Kathy tried again. "We're going to do the test on other children, and we'd like to know what you're afraid of, in case they are, too."

Corey looked miserable. "I don't know."

We walked back through the atrium of CHOP in silence, and then said goodbye.

On my way to my daughter Carly's apartment on the outskirts of the city, I mulled over the surprising end to the three-day visit. I didn't get it. I thought back to the other kids who'd been part of medical research in general, and the gene therapy saga in particular. How did they feel?

Did Sarah Nelmes, the milkmaid, mind when Edward Jenner took pus from the lesion on her hand in 1796, and did young James Phipps mind when Jenner scratched his arm and rubbed it with Sarah's secretions?

Who did David the bubble boy blame for his predicament?

How does Ashi feel today about being the first child to receive gene therapy?

Was Jesse Gelsinger's consent to gene therapy truly informed?

Jacob Sontag and Lindsey Karlin were babies when they first received gene therapy for Canavan disease directly into their brains, but they were old enough the second and third times to know what was coming. Were they afraid? Did they have a choice?

How much did Lori Sames tell Hannah when she took the little girl to the University of North Carolina for tests before the clinical trial for GAN? So far, Hannah thinks she only has floppy feet.

Maybe I think too much. When I told Carly about Corey's difficult day, ending with his refusal to participate in one final and important test, she was aghast—but not at his behavior. She was mad at me. "Well, of course he had a meltdown! The poor kid's sick of being a guinea pig!"

It turned out that Corey was, in fact, ill. Nancy posted on Facebook that Friday night that Corey got sick at a Rite Aid. I felt bad for misunderstanding him until my twenty-two-year-old daughter explained the situation from his point of view. But whether his tummyache was due to a virus, too many chocolate cupcakes, or fear of yet another uncomfortable medical test, his dramatic change in mood from one day to the next underscores the difficulty of planning a clinical trial that will include children.

Ten-year-olds don't think like adults. They can't ignore hurt to focus on a task that is boring at best, uncomfortable or painful at worst. Even if feeling well, can a ten-year-old be altruistic? When my doctor asked me, after plunging seven needles into my neck to pick off bits of my thyroid tumors, if I'd take another to provide a sample for a study, I didn't hesitate to say yes. But I was thirty-nine.

After two days of dark adapt/test/dark adapt/test, and the

stomachache and tears that Friday morning, could Corey even give informed consent? I think about what frightens me, and it's the feeling of being out of control. Following all those hours of being told what to do, was it so strange that when finally given a choice, Corey opted out?

The medical team had collected plenty of information from the days of testing to support what the family already knew—the gene therapy was a resounding success. The team evaluated Corey's test results throughout October and submitted them to the FDA. Fortunately, neither Jean Bennett nor Jennifer Wellman, who did the bulk of the work, needs much sleep, and prefer opposite hours. At scientific conferences they sometimes room together. Bennett sleeps from 9 p.m. to early morning, when she starts writing, while Wellman writes all night and then sleeps in. Once they'd finished the results report they hit the lecture circuit, speaking at several of the professional conferences that dot the fall calendars of biomedical researchers.

On October 28, 2010, the FDA contacted Jennifer Wellman. Enrollment could officially begin on treating the second eye.

20

THE FUTURE

IN HIS 2009 BOOK *OUTLIERS*, MALCOLM GLADWELL TALKS about the "10,000-hour rule": people who seem to achieve great success overnight have put in at least ten thousand hours practicing; he cites as examples the Beatles, Bill Gates, and Mozart. The rule is a variation of the breakthrough myth, for scientific advances and the development of new technologies do not happen overnight either, requiring two decades or more to get past the bumps and mature into familiarity. Test tube babies, human proteins made in bacteria, and pregnancy tests based on monoclonal antibody technology have all passed the milestone of acceptance.

Gene therapy has put in its requisite ten thousand hours, and then some. Begun officially in 1990, derailed in 1999, it is now back on track with the successful treatments of Leber congenital amaurosis type 2, adrenoleukodystrophy, and severe combined immune deficiencies. Gene therapy's time has finally come. The true beauty of this biotechnology is its broad applicability. Even though an inherited disease reflects malfunction of a particular gene and its protein, many other diseases that are acquired—not inherited—also stem from abnormal proteins. And not only are these conditions amenable to gene therapy too, but they are much more common than the single-gene diseases.

An example of a disease similar to Corey's, but not inherited and not nearly as rare, is age-related macular degeneration (AMD). It affects millions and is in line for eye gene therapy. Dick Breault's recent experience illustrates the evolution of treatment for AMD, from conventional drug to gene therapy.

The first and only sign that something was wrong with Dick's vision was the wiggly windows. It was August 2010, and he'd known that this could happen since his eye doctor had diagnosed the "dry" form of AMD in 2003. Back then it was hard to imagine that he could have a visual problem, but Dick had followed doctor's orders. "I took high-dose vitamin A and antioxidants, and my vision remained stable," he recalls. These pills would supposedly delay progression.

In 2007, Dick, a retired podiatrist, and his wife, Marlene, moved to a brand-new condo, just off the main road leading from Albany Airport to Schenectady, about an hour southeast of where Corey Haas lives. They had downsized from a house, but the condo had plenty of space—and windows. On that late summer day in 2010, Dick was looking out the large bay window onto the small back garden when he noticed waves in the pane. Suddenly remembering his diagnosis, he covered his right eye. No wavy wiggles. Then he covered his left eye, and there it was, a patch in the center of his vision that was changing, one moment moving, another blurry or blank. Alarmed, he went into his office and pulled out the Amsler grids that his ophthalmologist, Dr. Shalom Kieval, had given him to track the macular degeneration that he'd never actually experienced. The grids looked like graph paper, with a dot in the middle. When Dick stared at the dot with his right eye, as the doctor had instructed, the lines of the grid wavered. "So I called Dr. Kieval and he said to come in immediately." Marlene drove him to the doctor's office in Albany. There he got bad news and good news.

An arteriogram—X-rays of the blood vessels in his right eye—provided the bad news. Dick now had "wet" macular degeneration. Although 10 percent of people with the dry form develop the wet form too, the disorders are distinct, not a continuum. The chance of becoming completely blind from untreated wet macular degeneration is 85 percent.

The part of the eye affected in wet AMD is the thin layer next to the RPE, the part that in Corey's eyes can't utilize vitamin A. In Dick Breault's right eye, blood vessels from there were invading the RPE, snaking toward the rods and cones, oozing blood serum and swelling the delicate layers of the retina, like pulling apart the thin layers of a pastry. This was happening in the macula, the region at the back of the eye where the photoreceptors cluster, providing the most acute vision. While the RPE needs that blood supply to nourish the photoreceptors on its other side, too many blood vessels leak enough extra fluid to unhinge the precisely aligned rods and cones. Before Dick began to see the wavy windows, his RPE had started to vanish in spots, and that dying away had triggered the blood-vessel extension, a process called angiogenesis. Ironically, Dick had worked for several years in wound care, where the goal was to encourage growth of a blood supply into the feet. Now he had to squelch it.

The good news was that, like Corey Haas, Dick Breault was in the right place at the right time. "If you had come to me with this diagnosis in 2003, there wouldn't have been much I could do for you. But in 2006, that changed," said Dr. Kieval, referring to the drug Lucentis. "This drug is to ophthalmology what antibiotics were to infectious disease." Before Lucentis, Dick's good eye would have been affected within two years, and he would have slowly lost his vision.

The biotech giant Genentech manufactures Lucentis. The drug consists of a segment of an antibody molecule that binds certain receptors on the interiors of the tiniest blood vessels, preventing them from binding a protein called VEGF—for vascular endothelial growth factor—which extends blood vessels. A splash of VEGF onto the endothelium, which is the tilelike sheet of cells that folds to form a tubular capillary, adds more tiles to the ends and the tube grows. VEGF is a popular drug target because as a regulator of blood-vessel formation it is at the heart of many diseases. It's straightforward. To increase circulation in a crippled heart or to heal a wound, add VEGF. To dry up the blood vessels feeding a tumor or strangling an eye, block VEGF.

An antibody molecule is shaped like the letter Y. One arm of the top of the VEGF antibody was developed into the drug Lucentis.

The entire top of the Y was already on the market as the cancer drug Avastin, also from Genentech. Both blockbusters are "angiogenesis inhibitors," tailored to tackle VEGF. Both are injected, because antibodies are proteins, which fall apart in the stomach before they can be absorbed. But Lucentis, developed specifically for use in the eye, gets in and out of the body much faster than does Avastin.

Because Avastin dries up the blood vessels that feed tumors, its application to wet AMD was obvious, and many ophthalmologists prescribed it off-label before the FDA approved it for the eye disease in April 2011. It's much cheaper than Lucentis. At the time of this writing, a pharmacist can divide a dose of Avastin into several doses suitable for the eye, at about $50 a pop, compared to the $2,000 per dose of Lucentis. The first report from the CATT trial—Comparison of AMD Treatment Trials—found the drugs equally effective, either given monthly or "as needed." (Stay tuned as five additional clinical comparisons report in from Europe.)

More than fifty angiogenesis inhibitors are in clinical trials or are already on the market. They are the brainchild of Judah Folkman, MD, a Harvard medical researcher who first began thinking about the role of blood-vessel growth in cancer in the 1960s. In 1971 he published a seminal paper in *The New England Journal of Medicine* suggesting how tumors knit a blood supply to keep growing. But molecular biology hadn't yet shown how that might happen, and so like many ideas in biomedicine ahead of their time, Folkman's idea of angiogenesis in cancer was at best ignored, at worst dismissed. The research in his lab went on, however, as molecular biology flourished, and by the 1980s big pharma was paying close attention to developing a new class of drugs—angiogenesis inhibitors. Folkman passed away in 2008, living long enough to see his once radical idea restore vision to millions of people. At the American Academy of Ophthalmology meeting that year, he was hailed as a hero—but he certainly put in his ten thousand hours.

Lucentis is injected into the retina, once a month at first, and then gradually at longer intervals. Dick Breault got his first shot the day after the arteriogram—the wavy lines on his windows and the grid test meant he didn't have a moment to lose. Marlene helped him take numbing drops beginning the night before, and he had more drops right before the injection. By the second month, he could already read farther down on the eye chart using his bad eye. But Dick didn't need a visual acuity test to tell him the treatment was working—he could read the minuscule print in the phone book.

For Dick's third shot of Lucentis, I went along. He sat in the exam chair, and a technician leaned him backwards, then administered a series of eyedrops—numbing drops first, then two drops to dilate the pupil, a third drop of antibiotic that stings, and the fourth a numbing gel. The tech set out a little vial and a little needle, left for a few minutes as Dick's right eye lost sensation, then came back and painted the area around the eye with dark orange Betadine. She hurried off, leaving us to gaze at the poster of an eyeball on the back of the door. After another few moments, Dr. Kieval came in and got right down to business, preparing the injection. He swiveled on his stool to face Dick and smiled.

"How's your vision?"

The patient smiled back. "Gradually getting better. The grid is less distorted. I can read the phone book!"

"Fantastic! Now turn your head and open both eyes."

Dr. Kieval rubbed a cotton swab on the surface of Dick's right eye, looking for an area that was relatively free of blood vessels to minimize surface damage from the injection. He then held the eye open with a metal speculum as the technician applied more Betadine. Next, the doctor positioned a hollow blue tube over the center of the eye to guide and place the approaching needle.

"Now look to the left." Dick complied, and the doctor plunged the needle into the vitreous cavity of the right eye, withdrew it, then placed a cotton swab over the injection site. It took only a few seconds. Dr. Kieval removed the speculum and deftly irrigated the eye. "Now open your eyes wide."

Dick did so and visibly relaxed. It would be five weeks until the next injection, then six and then seven, until the shots come only once every three months. By the two-year mark, his vision might be completely restored, no further treatment necessary.

"Hey, back there," the doctor said to me. "Have you fainted yet? Actually, it's a pretty easy treatment to do."

He donned his ophthalmoscope and peered into Dick's eye. "It looks good in there! You're definitely getting better, we're documenting it."

The ophthalmologist swiveled around on his stool and looked at me, noticing my rapt attention. "Are you a medical student?"

When I told him about this book, and dropped a few scientific names, he grew so excited that he seemed to forget that he was supposed to be examining a patient into whose eyeball he had just injected what many in his field call a miracle drug.

"Jean Bennett! Al Maguire! I know them! I heard them speak just last weekend in Chicago, at the American College of Ophthalmology annual meeting. They showed the video of the boy, Corey, doing the obstacle course!"

I nodded and tried to speak, but he was too revved up to stop.

"I tell all my patients with genetic diseases of the eye about the LCA2 gene therapy. It's the proof of principle, and it's just a matter of time before it will be applied to other genetic diseases. It can be done and it works!"

Gene therapy for age-related macular degeneration was already well under way, and in just a few years, Dr. Kieval might be administering it himself. It's even more amazing than Corey's gene therapy for LCA2, because it has built-in controls. Jim Wilson and Jean Bennett are at the helm.

The first gene therapies treated rare inherited illnesses because their single-gene causes were well understood. The jump from Corey's disease, LCA2, to wet age-related macular degeneration elegantly demonstrates how a gene therapy strategy used to treat a very rare disease can be applied to a very common one.

Two million people have wet AMD in the United States alone.

That number is expected to increase by 50 percent as the population ages, with one in three people over the age of seventy-five affected. Wet AMD is the third most common cause of blindness worldwide, and the most common among the elderly and in developed nations. As a result, Lucentis has a market value of more than $2 billion a year. But the miracle drug doesn't work on everyone, and the frequent injections can raise the risk of infection, which can lead to bleeding and retinal detachment.

Gene therapy for wet AMD could render the blockbuster anti-angiogenesis drugs Lucentis and Avastin obsolete. Corey has shown the way—one shot should work. But the situation for AMD is much more complicated than that for LCA2.

Instead of a single mutation as the culprit, AMD springs from a tangle of contributing factors. It involves at least four genes related to immune system function, two related to cholesterol metabolism, and unknown environmental influences. And just as gene therapy for LCA2 wouldn't help someone with another form of the disease, a gene therapy that corrects one AMD gene wouldn't help a person with a mutation in a different gene. To get around this, gene therapy efforts for wet AMD center on the shared angiogenesis problem: blocking VEGF to stem the spurt of errant blood vessel growth.

Another challenge in designing gene therapy for wet AMD, compared with Corey's more tailored gene therapy, is that his just restores the eye's ability to use vitamin A. Too much of Corey's treatment probably wouldn't hurt. But if gene therapy for wet AMD works too well, blocking VEGF too much, part or all of the eye could lose its blood supply. If the loaded viral vectors should enter cells other than those to which they are targeted, as they did in Jesse Gelsinger's liver, they could interfere with circulation in other parts of the body. And should the gene therapy become toxic, the eye would have to be removed.

The first gene therapy clinical trial for wet AMD got under way in December 2009, at Johns Hopkins University and the University of Massachusetts Medical School, sponsored by Genzyme Corp. Four groups of participants, totaling thirty-four people over age fifty, are receiving 200 million to 20 billion "vector genomes," the metric of gene therapy. The vector is AAV2, the same subtype that's nestled forever in Corey's RPE. The ferried gene is part of the VEGF receptor, rather than the antibody part that is Lucentis. The idea is that if cells in the RPE crank out new receptors, they can sop up VEGF, keeping the VEGF from binding the real receptors, thereby stifling the blood vessel overgrowth. The Genzyme clinical trial should tell the researchers the most effective dose and how long the effect lasts. It's risky, though, because a too-high dose or delivery gone astray could dampen circulation elsewhere. That might happen once the treatment's approved, too.

A second gene therapy trial, from the Wilson-Bennett team, circumvents the potential problem of overdose by giving patients lifelong control over their gene therapy. Jim Wilson traces his inspiration for tackling AMD to an early interest in a similar condition that is twice as prevalent—diabetic retinopathy. Back in his training days at the University of Michigan, he became intrigued with diabetes because of how it deranges metabolism. Although gene therapy for diabetes itself would be too complicated because insulin levels are so tightly regulated, Wilson started to think about inventing a gene therapy to counter a *complication* of diabetes—diabetic retinopathy, which resembles AMD. The years of excess glucose in the blood damages blood vessels. In the front of the eye, the capillaries shred, triggering release of VEGF, which in turn stimulates new blood vessels to form. Their overgrowth continues and the unneeded blood vessels eventually snake back to infiltrate the delicate retina, where they swell and ooze serum.

To treat AMD (and perhaps one day diabetic retinopathy), a surgeon will inject AAV vectors carrying the DNA sequence for the part of the antibody molecule that is Lucentis, using basically the same procedure that Corey had. The vector will be the same, the cargo different, like switching passengers in a taxi. But since cells secrete the antibody segment, what if the intended effect spreads beyond the corrected cells?

The researchers are thinking ahead. "We include a switch to

turn the gene on or off, like a molecular rheostat," Wilson says. The switch is actually a small molecule, taken monthly or less frequently, that will counter the anti-VEGF activity to keep the effect at helpful, not harmful, levels. "Rather than repeat injections for years, you get one injection and you take pills or eyedrops" to control the activity, he adds. It's a wise marketing concept that makers of cars and computers might adopt—a built-in fail-safe mechanism that the operator controls, used only if needed.

A gene therapy that includes a control medication, taken noninvasively and only occasionally, is intriguing both scientifically and strategically. Drug companies likely won't back a forever fix unless there's also a forever market, or a onetime huge market, such as for a vaccine. And that may be why trying to find information about gene therapy efforts at large pharmaceutical companies is an exercise in frustration. Search for "gene therapy" on most pharma websites, and the computer gets hung up. Contact the media relations person, and the response is a variation on this sample missive:

> *——— does not currently have a gene therapy program. In the absence of clear and compelling evidence for a therapeutic benefit in an area with the potential to provide a reasonable return with minimal risk, there will be reluctance to invest considerable capital and resources.*

But don't Corey Haas and the others who've regained their vision, cognition, or immunity thanks to gene therapy provide that evidence? Perhaps the numbers simply need time to build.

A few companies are making an effort. GlaxoSmithKline got on board early when Amber Salzman learned her family had ALD, and the company is now developing gene therapy for half a dozen rare diseases with Fondazione Telethon and Fondazione San Raffaele in Italy. Sanofi-Aventis in 2009 partnered with Oxford BioMedica to develop LentiVector (aka HIV minus its bad genes) to target four eye diseases: wet AMD, Stargardt disease, Usher syndrome, and rejection of corneal transplants. Sanofi-Aventis also acquired the longtime biotech company Genzyme and its wet AMD program.

I can think of four reasons for big pharma to steer clear of gene therapy—and counterarguments.

1. *The diseases targeted so far are too rare.* Not so! Gene therapy has already gone from zebras to horses. Current applications target infectious diseases, heart failure, cancers, and the metabolic and genetic diseases on which the technology cut its teeth.
2. *It's still too soon.* Yes, about half the articles in *Human Gene Therapy* report experiments on rodents, rabbits, pigs, and monkeys. But starting with Corey's disease, gene therapy's time has come. The two-decade prelude to a breakthrough is up, while preclinical studies continue to refine today's gene therapies and fuel tomorrow's.
3. *Gene therapy doesn't work.* Yes, it does. Ask the parents of kids with ALD or SCID-X1 in Europe, or the young people with restored vision.
4. *Gene therapy works!* A forever fix will not produce a patient tied to a daily drug for a lifetime.

Could it be that the pharmaceutical industry is not in the business of *curing*, but of *treating*? This is why it took many years for ulcer treatment to change from a lifetime of acid-lowering pills to a forty-five-dollar, two-week course of antibiotics. A perusal of medical breakthroughs from the past half century reveals not a series of cures, but the creation of chronic conditions that lock in steady markets. Consider:

Depression: Popping a Paxil doesn't cure major depressive disorder. It takes weeks or months just to kick in, and then treatment lasts years.

Cancer maintenance: Gleevec, the lifesaving leukemia drug, is a daily pill, replaced with another when the cancer resists. Ditto Tamoxifen and Raloxifene for breast cancer. My tumor-ridden thyroid departed years ago, but I still take a tablet every day and always will.

Cardiac bypass: Yes, the circulation is rerouted, a heart salvaged, but a lifetime of statins, anticoagulants, and antihypertensives follows.

Transplants: Antirejection drugs are sometimes required for years.

So who's doing gene therapy?

Search clinicaltrials.gov for *gene transfer* and 320 or so hits come up. Most projects are sponsored by a combination of academic labs, children's hospitals, research institutions, and family-run not-for-profits. Barely one in five has a corporate connection, and nearly all of those are biotechs, not big pharma, or those describing themselves as a hybrid "biopharma."

When I was in grad school in the 1970s, newly minted PhDs who went into the fledgling biotech industry were seen as having gone over to the dark side. Twenty years later, Jim Wilson was raked over the coals for starting a company, if only a virtual one. Today commercialization is no longer a stigma, but a means to an end. Dr. Wilson is the scientific founder and is now a scientific advisor for ReGenX BioSciences, the Washington, D.C.–based company that is developing therapeutics as well as research tools based on novel forms of AAV.

But there is only so much that a biotech company can do. How can researchers passionate about gene therapy's promise convince pharma to come on board for the necessary phase 3 clinical trials? They need numbers to justify the investment and risk. And that's Jim Wilson's strategy.

Right now, gene therapy costs nearly $1 million per patient, counting the cost of vector development. This is comparable to the cost for certain transplants, but as Corey dramatically demonstrated, the hospital stay for gene therapy might be just days, compared to the three-week to three-month stay required for a stem cell transplant, for example. And costs must be weighed against long-term benefits. Corey's gene therapy removed the costs of lifelong frequent visits to the ophthalmologist and use of equipment to help him learn, not to mention providing the intangible benefits of restored vision. Such an economic analysis, however, would not apply to all applications of gene therapy equally. LCA2 would fare better than Canavan disease, for example, when longer survival extends expenses, and possibly suffering.

The current high cost of gene therapy will drop as vectors are tailored and stockpiled, and the technology is applied to more diseases, and that is where the numbers come in. It's already happening as clinical trials for macular degeneration and other eye diseases start. "We are moving from treating a rare genetic disease with a recessive deficiency to a platform for delivering therapeutic proteins. A patient can't swallow a protein drug. That's a real limitation to applying biologics. AAV is a way to deliver a therapeutic protein," argues Wilson. If gene therapy works on wet age-related macular degeneration, he says, the next step will be where he started—diabetic retinopathy.

Moving from LCA2 to AMD to diabetic retinopathy echoes the rationale underlying all gene therapy—what's good for the few will be good for the masses. Says Jean Bennett, "You have to walk before you can run. Taking a simple disorder, where we're certain it's one gene that lacks function, we can go to other single-gene disorders and then move to more common conditions."

Sometimes the evolution will be from the very rare to the merely rare, from the unicorns to the zebras. LCA2 gene therapy is already leading to gene therapies for other forms of Leber congenital amaurosis and to retinitis pigmentosa. The possibilities for the other diseases profiled in this book are even more profound.

First, think of similar conditions.

Ashi's disease, ADA deficiency, and David Vetter's disease, SCID-X1, both treatable with gene therapy, are immune deficiencies. *So is*

AIDS. Gene therapy for HIV infection is already happening. When a man in Germany was cured of AIDS after receiving a stem cell transplant in 2007 from a donor who had a mutation rendering his cells resistant to infection, the idea for HIV gene therapy was born. Unlike most such strategies, the one for HIV doesn't send in a gene but disables a gene, locking the virus out of our cells. So far, it works.

Adrenoleukodystrophy, the disease that claimed the lives of Lorenzo Odone and Oliver Lapin, is now treatable with gene therapy, thanks to the Salzman sisters. ALD is a demyelinating condition, stripping the fatty insulation off neurons. *So is multiple sclerosis.* The route of delivery in ALD, from the bone marrow to the brain, could be used to treat certain other brain disorders.

Treating Hannah Sames's giant axonal neuropathy with gene therapy in her spinal cord may pave the way for similar treatment of other motor neuron conditions. *These include spinal muscular atrophy (SMA), amyotrophic lateral sclerosis (ALS), and even spinal cord injuries.*

The Canavan kids—Lindsay, Jacob, Max, Lana, and the others—received gene therapy directly into their brains. *The same delivery system is being tested on patients with Parkinson's or Alzheimer's disease.*

Next, do the math.

Age-related macular degeneration = 5 million people, worldwide
Diabetic retinopathy = 100 million
HIV/AIDS = 35 million
Multiple sclerosis = 7 million
Spinal muscular atrophy = 700,000
ALS = 350,000
Spinal cord injury = 7 million
Parkinson's disease = 6 million
Alzheimer's disease = 35 million

Count rarer diseases too and the total easily exceeds 200 million people who might benefit from gene therapy—and that's only extrapolating from the small herd of zebras described in this book.

With the sequencing of the human genome and the twenty thousand or so potential gene therapy targets it includes, plus the hundreds of new varieties of viral vectors now available, gene therapy—alone or combined with stem cell therapy to better target and sustain it—is poised to explode in the future. It is the only way to correct the biochemical instructions that make us sick. Gene therapy will be a forever fix.

EPILOGUE

Corey has continued to make incredible progress since his first gene therapy. When I met him, he was still, fleetingly, hesitant in certain movements, and held his Nintendo right up to his face. But as the third anniversary of the gene therapy neared, his vision had vastly improved, so much so that he began spending much more time at the computer and reading on his Nook as he prepared for starting the sixth grade. He rode much farther on his bike, when not enjoying the family's new hot tub, and a high point of the summer was visiting the Great Escape, a nearby amusement park that years earlier had been strictly off-limits to the legally blind boy.

On September 1, 2011, Corey attended medical school, in a way. That morning, Dr. Jean Bennett was teaching a very special session of her genetics class. The 166 students had been in medical school a mere three weeks, and although they knew there would be two guests at the lecture, they were stunned enough to put down their smartphones when a big hairy dog came bounding in, dragging their diminutive professor by a leash.

It was Mercury, the dog who had cavorted with Corey at the Foundation for Retinal Research meeting the summer before, obviously still able to see well enough to rocket around the room, stopping occasionally to sniff and explore. Dr. Jean let him roam free as she prepared the class for the second guest, to whom they would ask questions. "Remember back to what it was like talking to a ten-year-old boy. They are often not very talkative, so you should think of leading questions." It was good advice.

Some of the students started scribbling questions while the lecture part of the session began. A PowerPoint presentation quickly covered the dog gene therapy, data from the clinical trials, and videos of Corey before and after treatment. Then the second guest appeared, via Skype.

"I'd like to introduce Corey Haas," said Dr. Jean from the lectern, as mouths dropped open and smiles broke out to match the one projected on the screen. Corey was speaking from his grandparents' home while Nancy and Ethan were at work, and his comments were to be a case presentation to launch the first set of lectures for the future doctors.

"What have you been up to, Corey? Doing anything exciting?" asked Dr. Jean.

"Well, I've been helping Grandpa with wood splitting, pulling out six-foot logs from the loader so they can get cut up."

After some more chitchat to catch up, Dr. Jean opened up forty minutes of questioning from the astonished students. At least thirty hands shot up.

"What did you notice that was different after the surgery? What could you do?"

"What do you want to be when you grow up?"

"Was it difficult deciding to be involved in this study?"

"Will the gene therapy be tested in even younger children?"

"How long do you think the effect will last?"

Corey's grandparents piped in when he wouldn't go much beyond yes or no, while Dr. Jean answered the technical questions. As the hour wound down, the students spontaneously rose and gave Corey a standing ovation, some moved to tears. And the boy who saved gene therapy was all smiles. Dr. Jean quieted the class and asked the final question:

"Corey, when would you like to have your second eye treated?"

"TOMORROW!"

NOTES

THE IDEA FOR THIS BOOK GREW OVER THE MANY YEARS OF WRITING a human genetics textbook and articles about genetics and biotechnology. The title comes from Lori Sames, who probably has no idea that she came up with it. We were perched on bar stools at a fund-raiser for Hannah's Hope, and we were watching people carrying Hannah around. Lori turned to me, tears in her eyes yet smiling, and said, "Ricki, if it works, it'll be a forever fix." When, a few months later, the researcher Paola Leone said much the same thing, I knew I had the title.

The Forever Fix wrote itself, the stories unfolding and interweaving seemingly on their own. Below are the major sources, both conversations and citations, used to propel the narrative.

PART I THE BEST THAT CAN HAPPEN

Corey Haas was in the right place at the right time.

1. Meeting Corey

Corey's story began, for me, in April 2008, when a few press releases stood out from the daily flood, written by Karen Kreeger from the University of Pennsylvania and Joey McCool Ryan from the Children's Hospital of Philadelphia. Their excellent reporting led to the medical journal papers about to be published, which led me to the research groups working on gene therapy for LCA2, Corey's disease—from the University of Florida, the Carver lab in Iowa, Moorfields Eye Hospital in London, and elsewhere. A few young adults had already regained vision, and I filed the reports away for updating my human genetics textbook.

Then an article about Corey appeared in the November 2, 2008, issue of *The Sunday*

Gazette, my hometown newspaper in Schenectady, New York, by Sara Foss. That went in the file too, and when my textbook was published a year later, Corey was in it, although I hadn't yet met him. I stared at his photo in my text. Was anyone writing a book about him? After all, his success had resurrected an essentially dead technology. I e-mailed Sara, who put me in touch with Ethan Haas, and to my astonishment, no, no one was writing about them. So I went to visit the family on December 5, 2009, and that is when and where the book opens.

2. The Road to a Diagnosis

The day I met Corey, Ethan and Nancy took turns detailing every step along the road to a diagnosis. Ethan shuffled copies of the entire medical record, recounting the visits to ever-farther-flung retinal experts until Dr. Anne Fulton made the critical connection to the dogs with LCA2. See H. Morimura et al., "Mutations in the *RPE65* Gene in Patients with Autosomal Recessive Retinitis Pigmentosa or Leber Congenital Amaurosis," *Proceedings of the National Academy of Science of the USA* 95:3088–93 (March 1998).

3. What's Wrong with Corey?

Talks at the Foundation for Retinal Research 6th Bi-annual LCA Family Conference at the University of Pennsylvania, July 31, 2010, explained the context of Corey's gene therapy. An early reference to Corey's disease is "A Case of Amaurosis after the Administration of Large Doses of Quinine-Recovery," D. B. St. John Roosa, in the *Transactions of the American Ophthalmologic Society* 4:431–34 (1887). The chapter ends with Francis Collins showing the Corey video to Congress to demonstrate the benefits of funding biotech research, but the NIH didn't fund his clinical trial; it funded the earlier trial at the University of Florida.

PART II THE WORST THAT CAN HAPPEN

The tragic tale of Jesse Gelsinger is easily reconstructed from the extensive media coverage, but talking to Paul Gelsinger, Dr. James Wilson, and others allowed me to see and tell the sad story from different perspectives.

4. The Breakthrough Myth

Theodore Friedmann met with me at the American Society of Gene and Cell Therapy (ASGCT) conference in Washington, D.C., in May 2010 and reviewed both the gene therapy timeline and the history of other technologies. A good reference is R. Powles, "50 Years of Allogeneic Bone-Marrow Transplantation," *The Lancet Oncology* 11:305–6 (April 2010).

Encyclopedias chronicle Edward Jenner's development of smallpox vaccine. The information on yellow fever is from *The American Plague* by Molly Caldwell Crosby (New York: Berkley Books, 2006). The Barney B. Clark papers, University of Utah, Special Collections, describe the artificial heart experiment. Also see G. J. Annas, "Law and the Life Sciences: Consent to the Artificial Heart: The Lion and the Crocodiles," *The Hastings Center Report* 13(2):20–22 (April 1983), and J. G. Copeland et al., "Cardiac Replacement with a Total Artificial Heart as a Bridge to Transplant," *New England Journal of Medicine* 351:859–67 (August 26, 2004).

"Baby Fae Stuns the World," wrote Claudia Wallis in the November 12, 1984, issue of *Time* magazine, while Charles Krauthammer analyzed the bioethics of using the baboon heart, "Essay: The Using of Baby Fae," in the December 3 issue.

5. Jesse and Jim

"Jesse's Intent" (www.circare.org), by Paul Gelsinger, provides his perspective on the 1999 gene therapy gone awry. See the National Urea Cycle Disorders Foundation website (www.nucdf.org) and "The Pediatrician's Guide to Ornithine Transcarbamylase Deficiency and Other Urea Cycle Disorders" from the National Organization for Rare Disorders (www.rarediseases.org) for information on Jesse's disease.

Dr. Wilson shared the story of Edwin, whose disease is described in William L. Nyhan and Dean F. Wong, "New Approaches to Understanding Lesch-Nyhan Disease," *New England Journal of Medicine* 334:1602–4 (June 13, 1996). A less technical article is "An Error in the Code," by Richard Preston, in *The New Yorker* (August 13, 2007). Dr. Wilson also shared his experiences with developing gene therapy for familial hypercholesteremia, cystic fibrosis, and OTC deficiency.

6. Tragedy

The Washington Post, *The Philadelphia Inquirer*, and *The New York Times* and others covered Jesse's gene therapy nearly daily. A comprehensive article is "The Biotech Death of Jesse Gelsinger," by Sheryl Gay Stolberg, *The New York Times Magazine* (November 28, 1999). Ten years later, on the date that his son died, Paul Gelsinger published "Seeking Justice for My Son" in *The Philadelphia Inquirer*. The archives of the Institute for Human Gene Therapy at Penn also cover the main events. Interviews with Dr. Wilson, Dr. Art Caplan, and the attorney Alan Milstein confirmed and supplemented the newspaper trail.

Technical papers highlight the search for what went wrong. The rationale for using disabled adenovirus as a vector is in G. P. Gao, Y. Yang, and J. M. Wilson, "Biology of Adenovirus Vectors with E1 and E4 Deletions for Liver-Directed Gene Therapy," *Journal of Virology* 70(12):8934–43 (1996). Three years later a paper indicates that monkeys can react adversely to AV in gene therapy: F. A. Nunes et al., "Gene Transfer into the Liver of Nonhuman Primates with E1-deleted Recombinant Adenoviral Vectors: Safety of Readministration," *Human Gene Therapy* 10(15):2515–26 (October 1999). The conclusions are in S. E. Raper et al., "Fatal Systemic Inflammatory Response Syndrome in an Ornithine Transcarbamylase Deficient Patient Following Adenoviral Gene Transfer," *Molecular Genetics and Metabolism* 80:148–58 (September–October), and Dr. Wilson's "Lessons Learned from the Gene Therapy Trial for Ornithine Transcarbamylase Deficiency," *Molecular Genetics and Metabolism* 96:151–57 (February 2009).

PART III EVOLUTION OF AN IDEA

Gene therapy has had many ups and downs.

7. The SCID Kids

Background on David Vetter is at www.scid.net/about.htm. See "The Boy in the Bubble" (www.pbs.org/wgbh/amex/bubble/peopleevents/p_vetter.html) and "Bursting the Bubble" (www.houstonpress.com/content/printVersion/218684), from April 10, 1997.

Early essays on gene therapy by scientific notables include "Dangerous Delinquents," by the Nobel laureate Joshua Lederberg (*The Washington Post,* January 8, 1967); "Gene Therapy for Human Genetic Disease?" by Theodore Friedmann and Richard Roblin, in *Science* 175:949–55 (March 3, 1972); and Dr. Friedmann's "Progress Toward Human Gene Therapy," in *Science* 244:1275–81 (June 16, 1989).

A pamphlet from the National Heart, Lung, and Blood Institute, "Curing Disease Through Human Gene Therapy," describes Ashi's experiment (www.thefreelibrary.com), which is reported in R. M. Blaese, "T Lymphocyte-Directed Gene Therapy for ADA-SCID: Initial Trial Results after 4 Years," *Science* 270:475–80 (October 20, 1995), and in C. A. Mullen et al., "Molecular Analysis of T Lymphocyte-Directed Gene Therapy for Adenosine Deaminase Deficiency: Long-Term Expression *in vivo* of Genes Introduced with a Retroviral Vector," *Human Gene Therapy* 7:1123–29 (June 10, 1996). Ronald Crystal published "Transfer of Genes to Humans: Early Lessons and Obstacles to Success" in *Science* 270:404–10 (October 20, 1995). I heard Dr. Don Kohn, director of the Human Gene Medicine program at UCLA, present his results at several meetings and interviewed him. See D. B. Kohn et al., "Engraftment of Gene-Modified Umbilical Cord Blood Cells in Neonates with Adenosine Deaminase Deficiency," *Nature Medicine* 1(10):1017–23 (October 1995).

An early paper by W. French Anderson is "Human Gene Therapy," *Science* 256:808–13 (May 8, 1992). *W. French Anderson: Father of Gene Therapy,* by Bob Burke and Barry Epperson (Oklahoma Heritage Foundation, 2003), details Dr. Anderson's early years. Jennifer Kahn's article in *Wired* magazine, issue 15.10, September 25, 2007, "Molest Conviction Unravels Gene Pioneer's Life," is a detailed look at the case.

8. Setbacks

Like the Jesse Gelsinger story, two other gene therapy tragedies were well reported in the general as well as scientific media. M. Cavazzana-Calvo et al., "Gene Therapy of Human Severe Combined Immunodeficiency (SCID)-X1 Disease," *Science* 288:669–72 (April 28, 2000), from Alain Fischer's group, presents initial evidence of efficacy ten months after treating the first three boys. French Anderson commented in "The Best of Times, the Worst of Times," on pp. 627–29 in the same issue. Further evidence for efficacy is in S. Hacein-Bey-Abina et al., "Sustained Correction of X-Linked Severe Combined Immunodeficiency by *ex Vivo* Gene Therapy," *New England Journal of Medicine* 346(16):1185–93 (April 18, 2002). A. Aiuti et al., "Correction of ADA-SCID by Stem Cell Gene Therapy Combined with Nonmyeloablative Conditioning," *Science* 296:2410–13 (June 28, 2002), is from the San Raffaele Telethon Institute for Gene Therapy in Milan, and detailed the protocol changes that Don Kohn and others needed to improve the safety of SCID-ADA deficiency gene therapy. S. Hacein-Bey-Abina et al., "A Serious Adverse Event after Successful Gene Therapy for X-linked Severe Combined Immunodeficiency," *New England Journal of Medicine* 348(3): 255–56 (January 16, 2003), reports the first leukemia case, with an editorial by Philip Noguchi, "Risks and Benefits of Gene Therapy," pp. 193–94.

News articles in *Science* by Eliot Marshall, Jocelyn Kaiser, and Jennifer Couzin followed the story: "Gene Therapy a Suspect in Leukemia-like Disease," October 4, 2002; "Second Child in French Trial Is Found to Have Leukemia," January 17, 2003; "Seeking the Cause of Induced Leukemias in X-SCID Trial," January 24, 2003; "As Gelsinger Case Ends, Gene Therapy Suffers Another Blow," February 18, 2005.

An excellent review article is D. B. Kohn et al., "Occurrence of Leukaemia Following Gene Therapy of X-Linked SCID," *Nature Reviews Cancer* 3:477–87 (July 2003). The technical report is S. Hacein-Bey-Abina et al., "*LMO2*-Associated Clonal T Cell Proliferation in Two Patients after Gene Therapy for SCID-X1," *Science* 302:415–19 (October 17, 2003), preceded by a "Perspectives" by David A. Williams and Christopher Baum, "Gene Therapy—New Challenges Ahead," pp. 400–401. U. P. Dave et al., in "Gene Therapy Insertional Mutagenesis Insights," *Science* 303:333–34 (January 16, 2004), suggests using a different virus. Alain Fischer explains what happened in "From Bench to Bedside" in the *European Molecular Biology Organization* 8(5): 429–32 (2007). A final report on the twenty patients in the SCID-X1 trial is in D. Kohn and F. Candotti's "Gene Therapy Fulfilling Its Promise," *New England Journal of Medicine* 360:518–21 (January 29, 2009). For a positive view, see the editorial "Gene Therapy Deserves a Fresh Chance," *Nature* 461:1173 (October 29, 2009).

The official explanation for Jolee Mohr's death is in K. M. Frank et al., "Investigation of the Cause of Death in a Gene-Therapy Trial," *New England Journal of Medicine* 361:161–69 (July 9, 2009). The attorney Alan Milstein blogged in "On Gene Therapy and Informed Consent," at blog.bioethics.net. Also see Art Caplan, "If It's Broken, Shouldn't It Be Fixed? Informed Consent and Initial Clinical Trials of Gene Therapy," *Human Gene Therapy* 19:5–6 (January 2008). The clinical trial protocol is number NCT00126724 at clinicaltrials.gov, "Study of Intra-articular Delivery of tgAAC94 in Inflammatory Arthritis Subjects," filed on August 2, 2005. The study wrapped up, as NCT00617032, on February 5, 2008. News releases from the FDA and Targeted Genetics flesh out the story.

9. Lorenzo and Oliver

I found the Salzman sisters in a November 6, 2009, article in *The Philadelphia Inquirer* by Tom Avril, "Sisters Rally Genetic Researchers to Tackle Fatal Brain Disease." We talked at the 2010 annual meeting of the ASGCT in Washington, D.C., and they provided much of the background and narrative for this chapter. I also spoke with Augusto Odone (Lorenzo's dad) and Jim Wilson, and heard Dr. Cartier speak at the meeting. For information on ALD see GeneReviews (genetests.org), the Stop ALD Foundation (www.stopald.com), and the Myelin Project (www.myelin.org).

Technical papers trace the evolution of ALD treatment from Lorenzo's oil to gene therapy. Reports from Patrick Aubourg and Nathalie Cartier go back to P. Aubourg et al., "A Two-Year Trial of Oleic and Erucic Acids ('Lorenzo's Oil') as Treatment for Adrenomyeloneuropathy," *New England Journal of Medicine* 329:745–52 (September 9, 1993). An editorial follows (pp. 801–2). Augusto and Michaela Odone challenge the 1993 paper's conclusion that the oil doesn't work in symptomatic patients in *New England Journal of Medicine*, pp. 1904–5 (June 30, 1994). Anne Hudson Jones analyzes the impact of the *Lorenzo's Oil* film in "Medicine and the Movies: Lorenzo's Oil at Century's End," *Annals of Internal Medicine* 133:568–71 (October 3, 2000).

Hugh Moser and colleagues, including Augusto Odone, support the oil in "Follow-up of 90 Asymptomatic Patients with Adrenoleukodystrophy Treated with Lorenzo's Oil," *Archives of Neurology* 62:1073–80 (July 2005). Also see "Lorenzo's Oil: Advances in the Treatment of Neurometabolic Disorders," by R. Ferri and P. F. Chance, pp. 1045–46. Dr. Moser et al. track the variability of the disease in 1,441 families in "Adrenoleukodystrophy: New Approaches to a Neurodegenerative Disease," *JAMA* 294:3131–34 (December 28, 2005).

The gene therapy paper is N. Cartier et al., "Hematopoietic Stem Cell Gene

Therapy with a Lentiviral Vector in X-Linked Adrenoleukodystrophy," *Science* 326:818–23 (November 6, 2009). A graphic is at www.aaas.org//news/releases/2009/1105sp_ald.shtml.

PART IV BEFORE GENE THERAPY

The Sames family is at the start of the gene therapy journey. Cari Scribner's article in my hometown newspaper introduced me to Lori and Hannah—"Rare Disease Afflicts Girl, Spurs Parents to Find Cure," *Daily Gazette* (May 13, 2008). For background on giant axonal neuropathy (GAN) see ghr.nlm.nih.gov/gene=gan and www.hannahshopefund.org/ and Genetics Home Reference. Most of the information in this part comes from time spent with Lori.

10. Hannah

Desperate parents seek practitioners of alternative medicine, and Chapter 10 opens with Lori and Hannah visiting a naturopath. See the National Center for Complementary and Alternative Medicine, nccam.nih.gov/health/naturopathy/ for a description of naturopathy.

GAN's rarity has placed it beneath the radar of the media, and so by necessity I used technical papers to trace the history. The earliest reports are B. O. Berg, S. H. Rosenberg, A. K. Asbury, "Giant Axonal Neuropathy," *Pediatrics* 49(6):894–98 (June 1972); S. Carpenter et al., "Giant Axonal Neuropathy: A Clinically and Morphologically Distinct Neurological Disease," *Archives of Neurology* 31(5):312–16 (November 1974); and J. G. Davenport et al., "'Giant Axonal Neuropathy' Caused by Industrial Chemicals," *Neurology* 26:919–23 (October 1976). A dog model is described in I. D. Duncan and I. R. Griffiths, "Peripheral Nervous System in a Case of Canine Giant Axonal Neuropathy," *Neuropathology and Applied Neurobiology* 5(1):25–39 (July 1978), and "Canine Giant Axonal Neuropathy; Some Aspects of Its Clinical, Pathological and Comparative Features," *Journal of Small Animal Practice* 22(8):491–501 (August 1981).

Cases accumulate in R. Tandan et al., "Childhood Giant Axonal Neuropathy: Case Report and Review of the Literature," *Journal of Neurological Sciences* 82:205–28 (1987), and M. Maia et al., "Giant Axonal Disease: Report of Three Cases and Review of the Literature," *Neuropediatrics* 19(1):10–15 (February 1988). The basis of the disease is explained in P. Bomont et al., "The Gene Encoding Gigaxonin, a New Member of the Cytoskeletal BTB/Kelch Repeat Family, Is Mutated in Giant Axonal Neuropathy," *Nature Genetics* 26:370–74 (November 2000), with the accompanying "Of Giant Axons and Curly Hair."

GAN mice are introduced in J. Ding et al., "Gene Targeting of GAN in Mouse Causes a Toxic Accumulation of Microtubule-Associated Protein 8 and Impaired Retrograde Axonal Transport," *Human Molecular Genetics* 15(9):1451–63 (2006), and more molecular details are in D. W. Cleveland et al., "Gigaxonin Controls Vimentin Organization Through a Tubulin Chaperone-Independent Pathway," *Human Molecular Genetics* 18(8):1384–94 (January 2009), and M. Tazir et al., "Phenotypic Variability in Giant Axonal Neuropathy," *Neuromuscular Disorders* 19:270–74 (February 2009). P. R. Lowenstein, "Crossing the Rubicon," *Nature Biotechnology* 27(1):42–44 (January 2009), suggests AAV9 can target motor neurons.

11. Lori

Lori and Matt turned to the Kellys (www.huntershope.org), Sharon Terry (www.pxe.org), the Miltos (www.nathansbattle.org), and the spinal muscular atrophy community (www.fightsma.org and www.smafoundation.org) for guidance in starting Hannah's Hope Fund and raising money and awareness. See A. MacKenzie, "Genetic Therapy for Spinal Muscular Atrophy," *Nature Biotechnology* 28(3):235–37 (March 2010) for background on SMA. Dr. Steve Gray, from the University of North Carolina, provided scientific details for developing gene therapy for GAN.

An early reference to misunderstanding of the drug approval process is R. Appelbaum et al., "The Therapeutic Misconception: Informed Consent in Psychiatric Research," *International Journal of Law and Psychiatry*, 5:319–29 (1982). The FDA Public Workshop on Cell and Gene Therapy: Clinical Trials in Pediatric Populations held in November 2010 provided background, as did Francis Collins's lecture at the 2010 American Society of Human Genetics meeting in Washington, D.C., a few days later. Also see D. Cressey, "Traditional Drug-Discovery Model Ripe for Reform," *Nature* 471:17–18 (March 3, 2011), and an editorial, "The Needs of the Few: Developing Drugs for Rare Diseases Is a Challenge That Requires New Regulatory Flexibility," *Nature* 466:160 (July 7, 2010), and E. Dolgin, "Big Pharma Moves from 'Blockbusters' to 'Niche Busters,'" *Nature Medicine* 16(8):837.

Other references about rare diseases are:

Office of Rare Diseases (NIH), rarediseases.info.nih.gov
Therapeutics for Rare or Neglected Diseases (TRND), www.genome.gov/27531965
NIH Bridging Interventional Development Gaps (BrIDGs), http://nctt.nih.gov/bridgs
Genetic Alliance, www.geneticalliance.org

PART V AFTER GENE THERAPY

What happens when gene therapy is only a partial forever fix? Consider Canavan disease.

12. Amazing Women

Doug Steinberg introduced me to the Karlins in "Gene Therapy Targets Canavan Disease," *The Scientist* (September 17, 2001). Conversations with the Karlins and Paola Leone comprise much of this chapter. To learn about Myrtelle Canavan see the Harvard Medical School Joint Committee on the Status of Women (www.hms.harvard.edu/jcsw/canavan.html).

13. The Jewish Genetic Diseases

A dated but comprehensive source is *Jewish Genetic Disorders: A Layman's Guide*, by Ernest L. Abel (McFarland & Company, 2001). Also see the National Foundation for Jewish Genetic Diseases (www.mazornet.com/genetics); National Tay-Sachs and Allied Diseases Association (www.ntsad.org), and the Jewish Genetic Diseases Consortium (www.jewishgeneticdiseases.org).

"The Rabbi's Dilemma," by Alison George, *New Scientist* 181:44–46 (February 14, 2004), discusses Rabbi Ekstein. See www.modernlab.org/doryeshirum.html for background on Dor Yeshorim.

Paola Leone asked me to include the story of Lana Swancey because non-Jews get Canavan disease too, and the association with one group can delay diagnosis for others. Michelle Swancey shared her family's experience.

14. The Patent Predicament

Canavan disease set the precedent for gene patenting, although the case against it was settled out of court. See *Greenberg et al. v. Miami Children's Hospital Research Institute* at www.kentlaw.edu/honorsscholars/projects/greenberg.html. Judith Tsipis, a plaintiff and head of genetic counseling at Brandeis University, shared her experience. The Canavan Foundation's press releases chronicled the case (www.canavanfoundation.org). Eliot Marshall covered it in "Families Sue Hospital, Scientist for Control of Canavan Gene," *Science* 290:1062 (November 10, 2000). Newspaper accounts include "Parents Suing over Patenting of Genetic Test," by Peter Gomer in *The Chicago Tribune* (November 19, 2000) and "COPING; A Postcard from Morgan's Twilight World," by Robert Lipsyte in *The New York Times* (April 14, 1996).

The discovery of the Canavan gene is reported in R. Kaul, G. P. Gao, K. Balamurugan, and Reuben Matalon, "Cloning of the Human Aspartoacylase cDNA and a Common Missense Mutation in Canavan Disease," *Nature Genetics* 5:118–23 (October 1993). Guangping Gao told me about it when I visited his lab.

Henrietta Lacks's cancer cells and John Moore's spleen are bioclassics. See *The Immortal Life of Henrietta Lacks* by Rebecca Skloot (Crown Publishing, 2010) and *Moore v. Regents of the University of California* (www.lawnix.com/cases/moore-regents-california.html).

15. Chasing Moonbeams

Many books, articles, and conversations contributed to this look at the ups and downs of planning a gene therapy trial, getting kids treated, and dealing with the aftermath. Lindsay Karlin's parents pioneered the gene therapy, and their openness led to many media reports. I interviewed Paola Leone, Ilyce Randell, Jordana Halovach (Jacob Sontag's mom), the Karlins, French Anderson, and Art Caplan. And I met Max at the thirteenth birthday party that no one thought he would see.

An early report on Canavan gene therapy is "Hope in a New Treatment for a Fatal Genetic Flaw," by Matthew Hay Brown in *The New York Times* (October 29, 1995). "New Zealand's Leap into Gene Therapy," by Eliot Marshall in *Science* 271:1489–90 (March 15, 1996), accuses Drs. During and Leone of skirting regulations. Dr. During answers in "Gene Therapy in New Zealand," *Science* 272:467–71 (April 26, 1996). "Gene Trial Causes Ethical Storm in New Zealand," by Sandra Coney in *The Lancet* 347:1759 (June 23, 1996), and Dr. During's response in the August 31 issue, "Gene Trial in New Zealand," *The Lancet* 348:618, continue the drama. Attorney Alan Milstein told me about a lawsuit against Dr. During's company Neurologix, Civil Action 11-10204-MLW.

"Fighting for Jacob" by Michael Winerip in *The New York Times Magazine* (December 6, 1998) is a yearlong look at the life of Jacob Sontag, with follow-up a decade later in "Taking a Chance on a Second Child." "Girl's Parents Plead for Gene Therapy to Resume" covers the Canavan kids' plight when gene therapy trials were suspended in the wake of the Gelsinger tragedy, in *WebMD Health News* (September 27, 2000). *His*

Brother's Keeper: One Family's Journey to the Edge of Medicine, by Jonathan Weiner (HarperCollins, 2004), details Matt During and Paola Leone's attempt to develop gene therapy for an ALS patient. It didn't work, but the book provides a nice history of the field.

The Karlins' story unfolded locally in "Gene Therapy Gives Girl Strength to Fight Disease," by Robert Miller in *The News-Times* (October 25, 2005), and "New Fairfield Teen Lives with Canavan Disease," by Sybil Blau, *The News-Times* (April 4, 2009). Also see *Lessons from Jacob*, by Ellen Schwartz with Edward Trapunski (Key Porter Books, 2006).

Technical articles are P. Leone et al., "Aspartoacylase Gene Transfer to the Mammalian Central Nervous System with Therapeutic Implications for Canavan Disease," *Annals of Neurology* 48(1):27–38 (July 2000), which details the gene therapy in New Zealand, and an editorial, D. J. Fink's "Gene Therapy for Canavan Disease?" on pp. 9–10. The clinical trial protocol is in C. Janson et al., "Gene Therapy of Canavan Disease: AAV-2 Vector for Neurosurgical Delivery of Aspartoacylase Gene (ASPA) to the Human Brain," *Human Gene Therapy* 13:1391–1412 (July 2002) and results in S. W. J. McPhee et al., "Immune Responses to AAV in a Phase 1 Study for Canavan Disease," *Journal of Gene Medicine* 8:577–88 (2006).

PART VI COREY'S STORY

16. Kristina's Dogs

LCA families, researchers, and Mercury the dog met at the 2010 Family Conference for the Foundation for Retinal Research, July 31.

"Talk to Kristina," said Dr. Jean mysteriously when I asked her who discovered LCA2 in dogs. Dr. Narfström happily helped. The first Briard paper is K. Narfström et al., "The Briard Dog: A New Animal Model of Congenital Stationary Night Blindness," *The British Journal of Ophthalmology* 73(9):750–56 (September 1989). Susan Semple-Rowland is another researcher who loves her experimental subjects, chickens. See M. Williams et al., "Lentiviral Expression of Retinal Guanylate Cyclase-1 (RetGC1) Restores Vision in an Avian Model of Childhood Blindness," *PLoS Medicine* 3(6):904–16 (June 2006).

17. Lancelot

Experiments on mammals other than humans precede clinical trials and continue during them. See M. Casal and M. Haskins, "Large Animal Models and Gene Therapy," *European Journal of Human Genetics* 14:266–72 (2006), and S. Cottet et al., "Biological Characterization of Gene Response in Rpe65$^{-/-}$ Mouse Model of Leber's Congenital Amaurosis During Progression of the Disease," *The FASEB Journal* 20:2036–49 (2006).

References for Dr. High's work on hemophilia B include N. C. Hasbrouck and K. A. High. "AAV-Mediated Gene Transfer for the Treatment of Hemophilia B: Problems and Prospects," *Gene Therapy* 15(11):870–75 (April 2008), and N. Boyce, "Trial Halted after Gene Shows up in Semen," *Nature* 414:677 (December 2001).

Early papers from the Narfström group include S. E. Nilsson et al., "Changes in the DC Electroretinogram in Briard Dogs with Hereditary Congenital Night Blindness and Partial Day Blindness," *Experimental Eye Research* 54(2):291–96 (February 1992); S. E. Wrigstad et al., "Ultrastructural Changes of the Retina and the Retinal Pigment Epithelium in Briard Dogs with Hereditary Congenital Night Blindness and

Partial Day Blindness," *Experimental Eye Research* 55(6):805–18 (December 1992); and A. Wrigstad et al., "Slowly Progressive Changes of the Retina and Retinal Pigment Epithelium in Briard Dogs with Hereditary Retinal Dystrophy," *Documenta Ophthalmologica* 87(4):337–54 (1994).

Papers in the October 1997 (vol. 17) *Nature Genetics* explain the LCA2-RPE65 link: A. F. Wright, "A Searchlight Through the Fog," pp. 132–34; F. Marlhens et al., "Mutations in *RPE65* Cause Leber's Congenital Amaurosis," pp. 139–41; and S. Gu et al., "Mutations in *RPE65* Cause Autosomal Recessive Childhood-Onset Severe Retinal Dystrophy," pp. 194–97. Two years later, Narfström's group described the dog mutation in A. Veske et al., "Retinal Dystrophy of Swedish Briard/Briard-Beagle Dogs Is Due to a 4-bp Deletion in *RPE65*," *Genomics* 57(1):57–61 (April 1999).

Details of LCA2 gene therapy in animal models are in L. Dudus et al., "Persistent Transgene Product in Retina, Optic Nerve and Brain after Intraocular Injection of rAAV," *Vision Research* 39:2545–53 (1999), and J. Bennett et al., "Stable Transgene Expression in Rod Photoreceptors after Recombinant Adeno-Associated Virus-Mediated Gene Transfer to Monkey Retina," *Proceedings of the National Academy of Science of the USA* 96:9920–25 (August 1999).

The surgery that let Lancelot see is described in G. M. Acland et al., "Gene Therapy Restores Vision in a Canine Model of Childhood Blindness," *Nature Genetics* 28:92–95 (May 2001). The question of who discovered what and when arises from an April 27, 2001, press release from Cornell University, which refers to the work in a paper that included Dr. Narfström as coauthor: "Congenital Stationary Night Blindness in the Dog: Common Mutation in the *RPE65* Gene Indicates Founder Effect," *Molecular Vision* 4:23–29 (1998). Dr. Narfström supplied the canines. The Cornell group's patent is number 6428958, "Identification of Congenital Stationary Night Blindness in Dogs," to develop carrier and diagnostic tests.

T. Michael Redmond's group at the National Eye Institute probed RPE65 protein from cows. See C. P. Hamel et al., "Molecular Cloning and Expression of *RPE65*, a Novel Retinal Pigment Epithelium-Specific Microsomal Protein That Is Post-transcriptionally Regulated *in Vitro*," *Journal of Biological Chemistry* 268(21):15751–57 (July 1993). They also discovered the role of RPE65 protein in utilizing vitamin A: T. M. Redmond et al., "*RPE65* Is Necessary for Production of 11-cis Vitamin A in the Retinal Visual Cycle," *Nature Genetics* 20:344–51 (1998). Dr. Redmond is also part of the team conducting Corey's clinical trial.

Papers following up LCA2 gene therapy are K. Narfström et al., "*In Vivo* Gene Therapy in Young and Adult *RPE65-/-* Dogs Produces Long-Term Visual Improvement," *Journal of Heredity* 94:31–37 (2003); K. Narfström et al., "Functional and Structural Recovery of the Retina after Gene Therapy in the *RPE65* Null Mutation Dog," *Investigative Ophthalmology and Visual Science* 44:1663–72 (2003); and K. Narfström et al., "Assessment of Structure and Function over a 3-Year Period after Gene Transfer in *RPE65-/-* Dogs," *Documenta Ophthalmologica* 111:39–48 (2005).

Researchers from several institutions collaborated for the dog trials. Papers include S. G. Jacobson et al., "Identifying Photoreceptors in Blind Eyes Caused by *RPE65* Mutations: Prerequisite for Human Gene Therapy Success," *Proceedings of the National Academy of Science of the USA* 102(17):6177–82 (April 26, 2005); G. M. Acland et al., "Long-Term Restoration of Rod and Cone Vision by Single Dose rAAV-Mediated Gene Transfer to the Retina in a Canine Model of Childhood Blindness," *Molecular Therapy* 12(6):1072–82 (December 2005); G. Le Meur et al., "Restoration of

Vision in *RPE65*-Deficient Briard Dogs Using an AAV Serotype 4 Vector That Specifically Targets the Retinal Pigmented Epithelium," *Gene Therapy* 14:292–303 (2007); and G. K. Aguirre et al., "Canine and Human Visual Cortex Intact and Responsive Despite Early Retinal Blindness from *RPE65* Mutation," *PLoS Medicine* 4(6):1117–28 (June 2007).

Work on dogs continues, in J. Bennicelli et al., "Reversal of Blindness in Animal Models of Leber Congenital Amaurosis Using Optimized AAV2-Mediated Gene Transfer," *Molecular Therapy* 16(3):458–65 (March 2008). D. Amado et al., "Safety and Efficacy of Subretinal Readministration of a Viral Vector in Large Animals to Treat Congenital Blindness," *Science Translational Medicine* 2(21):1–10 (2010), shows that treating a second eye is safe.

18. Success!

News releases hailed the first LCA2 gene therapy: "Results of World's First Gene Therapy for Inherited Blindness Show Sight Improvement," from the UCL Institute of Ophthalmology and Moorfields Eye Hospital NIHR Biomedical Research Centre (April 27, 2008); "Gene Therapy Improves Vision in Patients with Congenital Retinal Disease," from CHOP and the Penn School of Medicine (April 27, 2008); and "Safety Study Indicates Gene Therapy for Blindness Improves Vision," from the University of Florida, Gainesville.

Technical papers are J. W. B. Bainbridge et al., "Effect of Gene Therapy on Visual Function in Leber's Congenital Amaurosis," *New England Journal of Medicine* 358:2231–39 (May 22, 2008); A. M. Maguire et al., "Safety and Efficacy of Gene Transfer for Leber's Congenital Amaurosis," *New Engl. J. Med.* 358:2240–48 (May 22, 2008); A. V. Cideciyan et al., "Human Gene Therapy for RPE65 Isomerase Deficiency Activates the Retinoid Cycle of Vision but with Slow Rod Kinetics," *Proceedings of the National Academy of Science of the USA* 105(39):15112–17(September 30, 2008), and W. W. Hauswirth et al., "Treatment of Leber Congenital Amaurosis Due to *RPE65* Mutations by Ocular Subretinal Injection of Adeno-Associated Virus Gene Vector: Short-Term Results of a Phase 1 Trial," *Human Gene Therapy* 19:979–90 (October 2008).

Dale Turner's comments are courtesy of William Hauswirth. His group published an update in A.V. Cideciyan et al., "Vision 1 Year after Gene Therapy for Leber's Congenital Amaurosis," *New Engl. J. Med.* 361:725–27 (August 13, 2009) and again in S. G. Jacobson et al., "Gene Therapy for Leber Congenital Amaurosis Caused by *RPE65* Mutations," *Archives of Ophthalmology* doi:10.1000 (September 12, 2011). Other clinical trials include Hadassah Medical Organization (clinicaltrials.gov NCT00821340 phase 1) and Oregon Health and Science University and Applied Genetic Technologies Corp. (NCT00749957 phase 1/2). This research was featured in *The Seattle Times*, "Bellingham Brothers Get Experimental Gene Therapy in Attempt to Save Their Sight," September 7, 2010. Clinical trial NCT01208389 is for treating the second eye at CHOP.

Helpful documents from the NIH on clinical trial design are "Information for Institutions and Investigators Conducting Human Gene Transfer Trials" and "NIH Guidelines for Research Involving Recombinant DNA Molecules," aka "Appendix M." Reports on Corey's clinical trial are A. M. Maguire et al., "Age-Dependent Effects of RPE65 Gene Therapy for Leber's Congenital Amaurosis: A Phase 1 Dose-Escalation Trial," *The Lancet* 374:1597–1605 (November 7, 2009), and F. Simonelli et al., "Gene Therapy for Leber's Congenital Amaurosis Is Safe and Effective Through 1.5 Years after

Vector Administration," *Molecular Therapy* 18(3):643–50 (March 2010). A good review of everything is Artur V. Cideciyan, "Leber Congenital Amaurosis Due to *RPE65* Mutations and Its Treatment with Gene Therapy," *Progress in Retinal and Eye Research* 29:398–427 (2010).

19. Back to CHOP

For this chapter I followed Corey on his three-day, two-year follow-up to his gene therapy.

20. The Future

Outliers: The Story of Success by Malcolm Gladwell (Little, Brown, 2008) offers examples of successes that seemed overnight, but took thousands of hours—as is true for gene therapy.

References for age-related macular degeneration include N. Ferrara, "Vascular Endothelial Growth Factor and Age-Related Macular Degeneration: From Basic Science to Therapy," *Nature Medicine* 16(10):1107–11; M. J. Friedrich, "Seeing Is Believing," *JAMA* 304(14):1543–45 (October 13, 2010); K. Garber, "Biotech in a Blink," *Nature Biotechnology* 26(4):311–14 (April 2010); T. Hampton, "Genetic Research Provides Insights into Age-Related Macular Degeneration," *JAMA* 304(14):1541–43 (October 13, 2010); and J. T. Stout and P. J. Francis, "Surgical Approaches to Gene and Stem Cell Therapy for Retinal Disease," *Human Gene Therapy* 22:531–35 (May 2011) The statistics at the chapter's end come from a variety of sources.

ACKNOWLEDGMENTS

THE TALE OF GENE THERAPY COULD HAVE BEEN TOLD through any of many people's eyes, for any of many medical conditions. The first children to receive gene therapy were being treated right when I began the first edition of my human genetics textbook, and although I wasn't aware of it, I have been waiting since then for the perfect patient to help me tell the story. I needed an upbeat outcome, a condition affecting just one body part and caused by mutation in just one gene, and a person willing to share his or her experiences. I was lucky indeed to find the wonderful Haas family practically in my backyard.

Nancy, Ethan, and especially Corey, I can never thank you enough for the honor of telling your story. You have made medical history and given hope to so many. I thank Corey's ophthalmologist, Gregory Pinto, for sharing early observations, and David Harris at Boston Children's Hospital. This book would also not have been possible without the assistance of the research team: Jean Bennett, Al Maguire, and Jim Wilson at the University of Pennsylvania School of Medicine and Health System, and Kathleen Marshall and Katherine High at Children's Hospital of Philadelphia. Special thanks to Karen Kreeger at Penn and Joey McCool Ryan at CHOP, who facilitated access to the experts.

Corey was not the first and is no longer the youngest to have had gene therapy for Leber congenital amaurosis type 2 (LCA2). Several research groups have successfully performed the gene therapy. A big thank-you to William Hauswirth of the University of Florida for presenting the experiences of some of the first patients. A therapy becomes part of medical practice only after proving itself on many individuals, and each of the clinical trials for LCA2 is vitally important.

I thank Mary Dean, executive director of the American Society of Gene and Cell Therapy, for access to researchers at the 2010 annual meeting, and David and Betsy Brint, who invited me to the 2010 Family Conference for the Foundation for Retinal Research, where I met unforgettable families, including the Pletchers, Stevenses, and Coughlins. I hope your children will soon join Corey! I also thank the researchers who spoke at the meeting, and the star, Mercury, the part-Briard sheepdog who now has vision in both eyes.

Most new medical technologies take at least two decades to succeed, and by 2008 it was gene therapy's turn. But unlike other biotechnologies that proceed with incremental successes, gene therapy had to overcome a terrible tragedy—the death of eighteen-year-old Jesse Gelsinger. Many, many thanks to Jesse's father, Paul, for sharing his thoughts as well as resources, and to vectorologist Jim Wilson, attorney Alan Milstein, and bioethicist Art Caplan for providing their views. Thanks also to the science writers at *The Philadelphia Inquirer, The New York Times, The Washington Post, Science*, and *Nature* who followed in great detail the heartbreaking stories of Jesse as well as those of Jolee Mohr and David Vetter.

The idea of gene therapy was obvious once the genetic material had been identified and described, circa 1953. One of the first people to think and write about it was William French Anderson. I'd followed Anderson's career since the early days of gene therapy, and was thrilled when he agreed to assist me. I am grateful to French and Kathy Anderson for their insights and help. Allan Tobin, senior scientific advisor to the Cure Huntington's Disease Initiative and fellow textbook author, helped in recalling the dubious first attempts at gene therapy in the early 1980s. Thanks to Theodore Friedmann for telling me about the skepticism surrounding the first gene therapy, on Ashi DeSilva for ADA deficiency, and to Donald Kohn for providing background on the ADA deficiency and SCID-X1 clinical trials.

Corey's successful gene therapy is put into perspective through the lenses of similar attempts for three other single-gene diseases: adrenoleukodystrophy (ALD), giant axonal neuropathy (GAN), and Canavan disease. Any of these could easily fill the pages of a book. The ALD story reveals what parents go through when they desperately search for a treatment for a disease so rare that no one cares about it. GAN illustrates the "before" and Canavan disease the "after" of gene therapy. Thank you to the families and researchers who provided the backdrop to Corey's story.

Augusto Odone, Amber Salzman, Rachel Salzman, and Eve Lapin shared their experiences with ALD, and Nathalie Cartier and Patrick Aubourgh spoke about their development of the gene therapy at several meetings. Thank you to the inspiring and unforgettable Lori Sames and her family, Matt, Reagan, Madison, and of course Hannah. Thank you to Sarah LoBisco, the naturopath, for letting me sit in on her initial session with Hannah, to Steve Gray for developing gene therapy for this unicorn of a disease, to Anthony Brown for providing background on intermediate filaments, to Yanmin Yang for discussing her GAN mice, and to Christian Lorson and Karen Chen for including me in their discussion with Lori about similarities of GAN to spinal muscular atrophy.

A big thank-you to several families for the Canavan story: Helene, Roger, and Molly Karlin shared the experiences of their beloved Lindsay, the first to have gene therapy. Thank you to Ilyce Randell, who has been sharing her family's story with me since 2003, and Peggy, Mike, Alex, and Max. Jordana Halovach, Michelle Swancey, and Judith E. Tsipis, also thank you for sharing. Guangping Gao invited me to visit his lab, where he described the discovery of the Canavan gene and the patent controversy. A very special thank-you to the very special Paola Leone, who joins several other incredible women who form the backbone of this book. The families adore her, for good reason. I also thank Phil Milto and Sharon Terry for sharing their experiences with inherited diseases and parent activism.

The story of the Briard sheepdogs provided a bit of comic relief, yet also a serious look at competition in biomedical research. Many thanks to Kristina Narfström for telling me about her excitement in tracking down the *RPE65* mutation with her young students, and to Gillian McLellan for sorting out who was first to describe the dog mutation. Also many thanks to Susan Semple-Rowland for sharing the joys of working with chickens. Finally, thank you to Dick Breault for allowing me to accompany him to the ophthalmologist to get a shot in his eye, and to Dr. Shalom Kieval for indulging my interest.

The evolution of a narrative nonfiction book, especially one about a complex technology, requires many "readers." I had a great crew. From seemingly out of nowhere came Rita Ryan, science department chair at Cor Jesu Academy in St. Louis, who uses my human genetics textbook. We struck up a correspondence, I mentioned this book, and we both thought of the idea of using her high school students as readers. And they were all wonderful! A giant thank-you to Rita and her students Katherine Boul, Emma Gassett, Julie Schnur, Anna Scheuler, Lauren Straszacker, Zoe Preis, and especially Mary Haller, who will surely one day be a terrific editor. If any of you needs a college or grad school recommendation, please get in touch!

Many thanks to my friends who volunteered their time to help make this book a reality: John Davis, Jim Comly, Lucy Comly, Ana Pease, Sandra Latourelle, Mary Shannon, Hannah Valachovic, Sheila Winters, Tulle Hazelrigg, Lynn Leiberman, Lindsey Colman, Sue Conlin, Masako Yamada, Gautam Parthasarathy, Marlene Shaw, Marlene Breault, Annette Keen, Rachel Lieberman, Derek Johnson, Shelly Queneau Bosworth, Twitty Styles, Joan Gould, and Emmanuel Gokpolu. Thanks to Laura Newman for holding my hand in the world of social media and for publicity ideas and contacts. A very special thanks to my very best friend since we were thirteen, Wendy Josephs, who read every word of the very overwritten, never-ending first draft. Her optometric (eyeball) expertise was invaluable. She also took more than two hundred wonderful photographs of Corey Haas and Hannah Sames.

I am so thankful that Ellen Geiger, agent extraordinaire, was looking at her BlackBerry when I sent out my initial query shortly after meeting Corey in late 2009. Where other agents were "interested" in Corey's story, she was "excited." Publishing with St. Martin's Press fulfilled my dream since reading *And the Band Played On* by Randy Shilts in 1987. Thank you to editorial assistant Laura Chasen for keeping all the details straight and answering endless questions from a longtime textbook author venturing into a new area, and to Laura Clark and Nadea Mina. A huge thank-you to Nichole Argyres. I've been working with editors for many years, and learned much of what I know about writing from them, and Nichole is the absolute best. She is a master at stripping away words to reveal the shining story beneath.

Most important, thanks to my mother for taking an inquisitive little girl up to the very top floor of the American Museum of Natural History, demanding that a paleontologist identify the invertebrate fossils I had dug up on a weekend outing to tour the Baseball Hall of Fame that had veered into an excavation of a stream. My mother, Shirley Aaronson, saw the scientist in me and never for a nanosecond, back then in the 1960s, made me feel that a girl might face barriers in becoming a scientist. After I'd reached that goal, she filled the textbooks I'd written with yellow highlighter, ultimately the parts about cancer. I wish that she were here now to see the publication of *this* book, the book that I feel I was born to write.

A special thank-you to Marjorie Guthrie for introducing me to the rare disease community many years ago.

Finally, millions of thanks to my wonderful family, my husband, Larry, and daughters, Heather, Sarah, and Carly (and to all the cats and hippos and my tortoise, Speedy) for always being there to encourage me and to read multiple drafts of everything, and for learning to live with my constant uttering of "*Just one more sentence!*"